FLORIDA STATE
UNIVERSITY LIBRARIES

APR 4 1997

TALLAHASSEE, FLORIDA

ATLANTIC STUDIES ON SOCIETY IN CHANGE

NO. 85

Editor in Chief, Béla K. Király
Associate Editor in Chief, Peter Pastor

Demography of Contemporary Hungarian Society

Pál Péter Tóth and Emil Valkovics, Editors

Translated by Judit Zinner

Copy-edited by Professor Cecil D. Eby and Nóra Arató

Social Science Monographs, Boulder, Colorado
Atlantic Research and Publications, Inc.
Highland Lakes, New Jersey

Distributed by Columbia University Press, New York
1996

EAST EUROPEAN MONOGRAPHS, NO. CDLXI

The publication of this volume was made possible by grants from *Postabank és Takarékpénztár* (Postal and Savings Bank), *Nemzeti Kulturális Alap* (National Cultural Foundation), and *Országos Tudományos Kutatási Alap* (National Science Research Foundation), Budapest.

All rights reserved, which includes the right to reproduce this book or portions thereof in any form whatsoever, without the written permission of the publisher.

Copyright © 1996
by Atlantic Research and Publications, Inc.

Library of Congress Catalog Card Number 96-61475
ISBN 0-88033-358-8

Printed in the United States of America

Table of Contents

Table of Contents v
Preface to the Series and Acknowledgments vii
Emil Valkovics and Pál Péter Tóth
    Preface 1
Pál Péter Tóth
    Changes in the Number of the Hungarian Population
    Since the Conquest to Our Days 13
Rudolf Andorka
    Demographic Changes and Their Main Characteristics
    from 1960 to Our Days 21
Magdolna Csernák
    Marriage and Divorce in Hungary: Demographic
    Aspects of Changes 37
Ferenc Kamarás
    Birth Rates and Fertility in Hungary 55
Károly Miltényi
    Recent Changes in the Economic Activity and Retirement
    Patterns in Hungary: Financial and Social Implications . . 89
Péter Józan
    Changes in Mortality in Hungary between
    1980 and 1994 111
Pál Péter Tóth
    International Migration and Hungary 139

László Hablicsek
 Demographic Transitions and the Change of Regimes
 in Hungary 171
László Cseh-Szombathy
 The Role of Mental Elements in Demographic
 Phenomena 199
Marietta Pongrácz and Edit S. Molnár
 Adolescent Fertility 211
Emil Valkovics
 Cause of Death as a Factor in Creating Differences
 in Mortality Levels 231
List of Contributors 273
Volumes Published in "Atlantic Studies on Society in Change" . 275

Preface to the Series and Acknowledgments

The present volume is a component of a series which, when completed, will constitute a comprehensive survey of the many aspects of East Central European society. The books in the series deal with peoples whose homelands lie between the Germans to the west, the Russians, Ukranians and Belorussians to the east and north, and the Mediterranean and Adriatic seas to the south. They constitute a particular civilization, one that is at once an integral part of Europe, yet substantially different from the West. The area is characterized by a rich diversity of language, religion and government. The study of this complex area demands a multidisciplinary approach and, accordingly, our contributors to the series represent several academic disciplines. They have been drawn from the universities and other scholarly institutions in the United States and Western Europe, as well as East and Central Europe. More of the contributors of the present volume are scholars of the Hungarian Central Statistical Bureau – Demographic Institution, others are prominent Hungarian experts of the field.

The editors, of course, takes full responsibility for ensuring the comprehensiveness, cohesion, internal balance, and scholarly quality of the series he has launched. I cheerfully accept this responsibility and intend this work to be neither justification nor condemnation of the policies, attitudes, and activities of any persons involved. At the same time, because the contributors represent so many different disciplines, interpretations, and schools of thought, our policy in this, as in the past and future volumes, is to present their contributions without major modifications.

Budapest, October 23, 1996.

Béla K. Király
Editor in Chief

Emil Valkovics and Pál Péter Tóth

Preface

The essays in this volume deal with the most important demographic problems of Hungary. On January 1, 1996 Hungary had 10,214,000 inhabitants on 93,011 square kilometers. The size of the population and the territory of the country have undergone significant changes in the past seventy years and these changes have influenced the age distribution and other demographic characteristics of the population. It may, therefore, be important to touch upon these events and their consequences in a few sentences.

While up to 1918 Hungary was the sixth largest country in Europe by area, even larger than Great Britain and Italy, at the end of the war it became one of the small countries of Central Europe. Owing to the peace treaty closing the war it lost 63.5 per cent of its population and 71.4 per cent of its former territory. Instead of the 20,886,487 inhabitants registered by the census of 1910, it had than 7,615,117, and instead of its former 324,411 square kilometers it had a territory of 93,073. Hungary became a homogeneous country with small minority populations. At the same time more than three million ethnic Hungarians became citizens of Austria, Romania (whose territory was greatly extended by the peace treaty), and the two newly formed countries of Czechoslovakia and the Serbian-Croatian-Slovene Kingdom. This situation changed a bit after 1938 when the First and the Second Vienna Awards, and the reannexation of Sub-Carpathia and former Southern Hungary increased the population by nearly four and a half million as compared to the situation in 1920. The territorial gain was 79,131 square kilometers. This was, however, a very short-lived situation, since the Paris Peace Treaty of 1947 closing the Second World War restored the Hungarian borders valid prior to the war, and even gave three more villages to Czechoslovakia. As a result of the repeated border changes, the human losses during the war, the refugees fleeing the Soviet army occupying the country, the compulsory deportations, and the population exchange between Hungary and Czechoslovakia the population of Hungary on January 1, 1949 amounted only to 9,204,799 inhabitants. In the following decade and a half the demographic situation of the country was steadily improving, the number of the population was quickly increasing, and fertility rates were promising. At the same time, mortality was getting lower. More than twice as many marriages were registered till the middle of the 1970s than today, and the number of divorces during the years immediately following the end of World War II was less than one third of the present figures. Families had more members, and fewer people lived alone without a family than today. Also the number and rate of one-parent families was low. Although the aging tendency of the population was a problem, its social and economic consequences did not seem to be menacing or unmanageable. Emigration and immigration did not have a significant demographic effects until 1956, when nearly 200,000 people left the country. This loss could, however, be made up relatively quickly, partly by means of immigration. All in all, the population of Hungary was steadily growing until 1979 when it was above 10,700,000. However, from that time on, mortality superseded the rate of live births and, including also the legal and illegal immigration, the natural decrease of the population became a steady feature of Hungarian population development.

Owing to the various causes and their interactions, the demographic situation of the country can be called critical at the moment. The most serious problems are the following:

- the decrease of the number of the population,
- the fall of fertility level below the level necessary for the simple reproduction of the population,
- the growing mortality level,
- the significant aging of the population with all its social and economic consequences,
- the decrease of the number of marriages,
- the increase of the number of divorces,
- the decrease of the number and size of the families,
- the increase of the number and rate of cohabitation,
- the increase of the number and rate of one-parent families,
- international migration as a significant factor in Hungarian population development, and the growing losses owing to emigration that cannot be made up for any longer.

The essays of the book analyze the most serious problems. The following data are meant to support their analysis.

Shrinking of the Hungarian Population

The rate of natural population increase on the present-day Hungarian territories was positive for a long time after 1876 with the exception of a few years only. The exceptions were the years of the First World War (1915, 1916, 1917, 1918), and the last year of the Second World War (1945). However, since 1981 this rate has already been negative. Calculated for thousand inhabitants the figures are the following:[1]

1981	-0.2 (per thousand)
1982	-1.0
1983	-2.0
1984	-2.0
1985	-1.6
1986	-1.8
1987	-1.6
1988	-1.5
1989	-2.1
1990	-1.9
1991	-1.7
1992	-2.6
1993	-3.2
1994	-3.0

According to population projections, the natural decrease of the Hungarian population is likely to continue in the future. The male population is diminishing faster than the female one. In 1993, for example, the decrease was 4.1 per thousand among men and 2.4 per thousand among women, which meant a decrease of 3.2 per thousand all in all. The so-called intrinsic rate of population growth would mean a still faster decrease. While the actual decrease in 1993 was 3.2 per thousand, the intrinsic rate of population growth calculated on the basis of the 1993 data would be -8.2 per thousand, i. e., this would be the rate of the population decrease, should the general age-specific fertility and mortality rates be unchanged for such a long period that the population would become stable.

Falling Fertility Level

The level of fertility fell so steeply as early as the late 1950s and early 1960s that it was no longer enough for the simple reproduction of the population. The only period when the situation was a bit more favorable was between 1974 and 1978. In all the other years the gross reproduction rate was below 1.00, so the simple reproduction of the population would not be ensured even if no women died until the age of fifty. The changes of the total fertility rate and the gross and net reproduction rates are shown in the following chart:

Changes of the total fertility rate* and the reproduction rates** [2]

Year (average of years)	Total fertility rate	Gross	Net
		reproduction rate	
1921	3.80	1.828	1.128
1930–1931	2.84	1.385	1.010
1940–1941	2.48	1.194	0.972
1949	2.54	1.223	1.060
1950	2.62	1.259	1.106
1960	2.02	0.975	0.917
1970	1.97	0.953	0.912
1971	1.82	0.931	0.891
1972	1.93	0.931	0.894
1973	1.95	0.943	0.906
1974	2.30	1.117	1.069
1975	2.38	1.157	1.111
1976	2.26	1.096	1.056

* The total fertility rate expresses the possible number of children born to a mother by the fertility rates of a given year.
** The gross reproduction rate shows the possible number of daughters born to mothers during their lifetimes by the fertility rates of a given year. The net reproduction rate shows the number of daughters born to a mother reaching their productive years by the mortality rates of the same year. If the value of the latter is 1.00, the population is stagnating. Values above 1.00 mean increase, and values below 1.00 mean decrease in the given population.

Changes of the total fertility rate and the reproduction rates
(continued)

Year (average of years)	Total fertility rate	Gross	Net
		reproduction rate	
1977	2.17	1.056	1.021
1978	2.08	1.010	0.979
1979	2.02	0.985	0.956
1980	1.92	0.937	0.909
1981	1.88	0.919	0.894
1982	1.78	0.865	0.843
1983	1.73	0.859	0.837
1984	1.73	0.849	0.826
1985	1.83	0.892	0.867
1986	1.83	0.893	0.870
1987	1.81	0.880	0.859
1988	1.79	0.872	0.852
1989	1.78	0.852	0.831
1990	1.84	0.906	0.889
1991	1.86	0.906	0.885
1992	1.77	0.858	0.839
1993	1.69	0.821	0.804
1994	1.64	0.735	0.721

Increasing Mortality Level

The mortality level of the Hungarian population was constantly improving in the first twenty years after the Second World War, then they began to rise in the case of men and stagnate in the case of women (or improved only slightly). Male mortality, increasing since the mid-1960s, started to deteriorate still faster in the first half of the 1990s. Statistics of mortality[3] show the average life expectancy at birth as follows:

Year	Males	Females
1965	66.71	71.54
1970	66.31	72.08
1971	66.11	72.04
1972	66.85	72.57
1973	66.65	72.49
1974	66.52	72.38
1975	66.29	72.42
1976	66.60	72.50
1977	66.67	72.99
1978	66.08	72.74
1979	66.12	73.03
1980	65.45	72.70
1981	65.46	72.86
1982	65.63	73.18
1983	65.08	72.99
1984	65.05	73.16
1985	65.09	73.07
1986	65.30	73.21
1987	65.67	73.74
1988	66.16	74.03
1989	65.44	73.79
1990	65.13	73.71
1991	65.02	73.83
1992	64.55	73.73
1993	64.53	73.81
1994	64.84	74.23

Demographic Aging of the Population

The Hungarian population is constantly aging, though not at an even pace and not without interruptions. The mean age of the population is rising, and the rate of people at and above sixty is growing[4].

Year	Mean age			Share people above 60 (per cent)		
	males	females	both	males	females	both
1869	25.9	25.2	25.5	5.4	4.8	5.1
1880	26.3	26.2	26.3	6.7	6.7	6.7
1890	26.4	26.4	26.4	6.7	7.1	6.9
1900	26.9	26.9	26.9	7.3	7.7	7.5
1910	27.2	27.3	27.2	7.8	8.2	8.0
1920	28.5	28.9	28.7	9.0	9.0	9.0
1930	29.3	30.2	29.8	9.7	9.9	9.8
1941	31.0	32.1	31.6	10.2	11.2	10.7
1949	31.5	33.3	32.4	10.7	12.6	11.6
1960	32.5	34.8	33.6	12.3	15.2	13.8
1970	34.3	37.0	35.7	15.1	18.9	17.1
1980	34.6	37.7	36.2	14.6	19.4	17.1
1990	35.5	39.0	37.3	15.8	21.8	18.9
1994	–	–	–	16.0	22.4	19.3

Note: Up to 1930 the data refer to December 31, from 1941 they refer to January 1 each year. The tendency of aging will continue in the future, too, even if a bit slower sometimes.

The decreasing number of marriages[5]

Year	Number of marriages per thousand unmarried	
	men	women
	aged 15 years and above	
1941	62.9	54.3
1948	...	...
1960	88.6	63.5
1970	81.7	62.1
1980	68.7	51.1
1988	48.8	37.0
1989	48.3	36.8
1990	47.4	35.9
1991	42.3	32.3
1992	37.3	28.7
1993	35.4	27.2
1994	34.6	26.7

This tendency equally applies to first marriages and remarriages. Even within the statistics for cohorts of first marriages the share of married people is falling and that of singles is growing.

Increase in the number of divorces[6]

Years (average of years)	Number of divorces per thousand	
	existing marriages	newly registered marriages
1930–1931	2.6	63.9
1938	2.8	77.5
1948	5.3	113.2
1960	6.5	187.3
1970	8.4	236.4
1980	9.9	346.0
1988	9.1	362.1
1989	9.6	372.7
1990	9.9	374.8
1991	9.8	399.2
1992	8.8	379.0
1993	9.2	413.1
1994	9.8	432.7

The rate of marriages ending in divorce is growing also within the marriage cohorts.

The growing rate and number of one-parent families[7]

Year	One-parent families		
	Total	Father with children	Mother with children
1960	368,584	32,997	335,587
1970	293,451	37,365	256,086
1980	341,227	56,045	285,182
1990	449,862	89,125	360,737
	1960 = 100		
1960	100.0	100.0	100.0
1970	79.6	113.2	76.3
1980	92.6	169.8	85.0
1990	122.1	270.1	107.5
	Per cent of all families		
1960	13.4	1.2	12.2
1970	10.2	1.3	8.9
1980	11.3	1.9	9.4
1990	15.5	3.1	12.5
	Per cent of families with children		
1960	19.3	1.7	17.6
1970	15.3	1.9	13.4
1980	17.4	2.9	14.5
1990	23.6	4.7	18.9

International migration has been a factor in Hungarian society in new ways since the change of political regimes. Until 1990 it was very difficult for a foreigner to stay in Hungary as an immigrant except for cases of family unification. Changes began in 1988 when more than 13,000 Rumanian citizens refused to return to their country. Subsequently, Hungary joined with the international agencies in regulating the legal status of refugees, immigrants and emigrants alike. Since 1990, more than 130,000 refugees arrived first from Romania, then from the countries of former Yugoslavia after the outbreak of the war. At the same time, also more than 100,000 immigrants arrived. Nearly 55,000 people were granted Hungarian citizenship, most of whom were ethnic Hungarians from the neighboring countries. Since the registration duty has been cancelled, nobody knows the exact number of those who left the country in the past five years, the assessed number is nearly 100,000.

As statistical data for the first half of 1996 reveal, the situation outlined above and analyzed in the essays of the present volume is further deteriorating. In the first quarter of the year, the Hungarian population declined 14,000. This is an unprecedented pace of decrease when compared with the same period in the preceding years. Should this tendency remain unchanged for the rest of the year, the natural decrease will be much above 50,000 for 1996. According to the 1990 census results, the population of Hungary was 10,375,000 in that year. Between 1990 and 1993 the number of the population decreased steadily. On January 1, 1993 it was 10,310,000, and it is 10,200,000 at present. According to calculations for the future, decrease between 1993 and 2000 will be 142,000, between 1993 and 2010 it will be more than 400,000, and between 1993 and 2020 nearly 820,000. So the number of the Hungarian population in 2020 is expected to be 9,500,000.

Long-range population projections speak of about 150,000 deaths yearly, while the number of births will continue to decrease significantly. The natural decrease of the population (the difference of the number of deaths and births) will be above 40,000 per year from 2010 onward to an ever growing degree. At the same time, a slight increase in the number of births can be expected in the 1990s from the more numerous generations of women born in the 1970s. Aging of the population will also continue, as the rate of people under forty will decrease, and that of those above forty will increase in the future.

According to László Hablicsek: "Low birth rates and surplus deaths lead us to the conclusion that dependents, both young and old, will hopefully mean less burden for a certain period, should other factors not intervene. On the other hand, the expected natural loss of 40,000 per year after 2010 indicates that the decrease of the population will gradually become an irreversible and self-inducing process if the demographic characteristics of 1992 remain unchanged in the future."[8]

Notes

1. *Demográfiai Évkönyv* (Yearbook of Demography), (1993), 38–39.
2. Ibid., 414.
3. Ibid., 414.
4. Ibid., 22–25.
5. Ibid., 50.
6. Ibid., 80.
7. Klára Serfőző-Szukics, *Az egyszülős családok társadalmi-demográfiai jellemzői* (Social and Demographic Characteristics of One-Parent Families), KSH NKI Kutatási Jelentései (Research Reports of the Demographic Research Institute of the Central Statistical Office), no. 55 (1995), 11.
8. László Hablicsek, *Az első és második demográfiai átmenet Magyarországon és Közép-Kelet-Európában* (The First and the Second Demographic Transition in Hungary and in East-Central Europe), KSH NKI Kutatási Jelentései, no. 54 (1995), 60.

Pál Péter Tóth

Changes in the Number of the Hungarian Population Since the Conquest to Our Days

The ancient Hungarians arrived at the territory of their present-day homeland in the Carpathian Basin in 896, from a territory called Levedia, beyond the Dnieper River and passed along the Lower Danube. At that time there was no unified political power in the region to resist them, so they occupied first the Great Plains and Transylvania, then took the Transdanubian region from the Franks, and finally the western part of later Upper Hungary from the Moravians without serious resistance. The scattered population of the occupied territories consisted of Avars, Slavs, and some Bulghars. By 907 the Hungarian conquest was an accomplished fact. In 970, when Géza was elected ruling prince of the Hungarians, new centralizing processes came to be dominant in the society. As a result, the tribes and the clans gradually lost their independence. Prince Géza established relations with his former western enemies, asked for and obtained Christian missionaries from the German King Otto. Géza's son, István (later Stephan I) assumed power in 997 and completed the organization of the Hungarian state. He forced his people to adopt Christianity and ordered that every ten villages should build a church and keep a priest. The efforts of Stephan in creating the feudal Hungarian state were rewarded by Pope Silvester II, who sent him a crown with which he was invested in 1001 "Apostolic King" of Hungary. By this the rule of the House of Árpád (the forefather who was ruling prince at the time of the Conquest) began and lasted until 1301, the death of Andrew III.[1]

For want of written sources, the strength of the Hungarian people at the time of the Conquest can only be estimated, and it is thought to be around 400,000–500,000. In the following two centuries they absorbed another 200,000 people of foreign origin, while part of the population found in the Carpathian Basin, especially the Slavs in the valleys of the Carpathians, were undisturbed. In the 11th and 12th centuries the Hungarian kings settled Pechenegs and Uzhes in the country, as a result of which the number of the population rose probably to around 1.8–2.2 million by the end of the 12th century. The Mongol invasion had a devastating effect on the Hungarian population. Although the Mongols stayed in Hungarian territory only for two years (1241–1242), the devastion was significant. The destruction of the village structure built up by then shows that 20 per cent of Western Hungary and 60 per cent of Eastern Hungary (especially of the settlements on the Great Plains) were destroyed. To replace the population loss, the king invited German, Walloon, Italian, and other settlers to live in Hungary. It was in those years that the nomadic Cumanians of Turkish origin were settled in the central territories.[2]

Owing to wars, epidemics, and famine the number of the population rose only slowly after the devastating Mongol invasion. It was not until late in the mid-14th century that the population reached 2,000,000 again. As a result of a relatively steady growth in the following two centuries, it was around 3.5–4 million by the early 16th century. The next two centuries of Turkish occupation, however, proved to be a catastrophe again. So by the early 18th century the Hungarian population did not exceed 3.5 million, and even its composition had changed greatly. While in the 16th century about 80 per cent of the population spoke Hungarian as a mother tongue, by the 18th century the rate declined to about 35–45 per cent. These changes in the language structure were partly due to human losses and the subsequent settlement of foreigners during the previous centuries, and partly to the fact that while the destruction of the central territories inhabited mostly by ethnic Hungarians was significant, the Slovaks and Romanians living in the peripheries increased in number, and a great number of Croats and Serbs arrived because of Turkish pressure.[3] Also Germans were settled in the depopulated regions from the 17th century onwards. These were the changes influencing the ethnic distribution of the population up to our present day. The census executed during the reign of Joseph II in the 1780s found about 9,515,000 people living in Hungary. The number of ethnic Hungarians could be 3–3.5 million, which was nearly 38 per cent of the total population. Between the late 1700s and 1850 there was a gradual increase, then from the mid-19th century a more rapid one in the number of the population, so the population increased by 37 per cent (more than 5,600,000 people) in a century (Croatia and Fiume not included). The increase was significant in spite of the three cholera epidemics in the 1800s, the last of which caused the death of 180,000 people in 1872–74.[4] In 1910 the population of Hungary exceeded 18 million (Croatia not included). This change was closely related to the bourgeois development of the society, as a result of which mortality decreased much more rapidly than fertility, in contrast with the definite tendency of high mortality and fertility in the 19th century. Under the new circumstances between the 1880s and the beginning of the First World War the yearly average of natural increase was 11 per thousand in spite of the emigration of 1.2 million people, mostly to the United States. The following table illustrates the distribution of the population of the Hungarian Kingdom by nationality on the basis of three different census tabulations.[5]

Nationality	1800		1900		1910	
	persons	%	persons	%	person	%
Hungarian	6,404,070	46.6	8,651,520	51.4	9,944,627	54.5
German	1,870,272	13.6	1,999,060	11.9	1,903,357	10.2
Slovak	1,855,451	13.5	2,002,156	11.9	1,946,357	10.7
Romanian	2,403,041	17.5	2,798,559	16.6	2,948,186	16.1
Ruthene	353,229	2.6	424,774	2.5	464,270	2.5
Croat			196,781	1.2	198,700	1.1
Serb	639,986	4.6	520,440	3.1	545,833	3.0
Other	223,054	1.6	244,956	1.4	313,203	1.7
Total	13,749,603	100.0	16,838,255	100.0	18,264,533	100.0

In 1910 the Hungarian Kingdom had 18,264,533 inhabitants. The ones whose mother tongue was Hungarian amounted to nearly 55 per cent of the total, the most numerous nationalities being the Romanians with 16.1 per cent, the Slovaks with 10.7 per cent, and the Germans with 10.4 per cent. There were also 464,270 Ruthenes, 461,516 Serbs, 282,653 Croats, 108,825 Gipsies, 77,398 Slovenes, 38,225 Poles, 22,924 Bulgarians, and 168 Greeks living within Greater Hungary. (Statistics of the day do not speak of any Ukrainians.) All the rest amounted to 65,906 people. This means that except for the Romanians, the Slovaks, and the Germans, the rate of all the other nationalities was insignificant in Hungary in 1910.

This situation was drastically changed by the peace treaty closing the First World War. The victorious powers put an end to the multinational Hungarian Kingdom and brought about three multinational states in the region, namely, Czechoslovakia, Romania, and the Serbian-Croatian- Slovene Kingdom, two of which (Czechoslovakia and Yugoslavia) have recently disintegrated. As a result of the Great Powers' decision Hungary lost 71.4 per cent of its territory and 65.6 per cent of its population. Its territory decreased from the former 325,411 square kilometers to 93,073 square kilometers, and its population dropped from 18,262,533 to 7,615,117.

A major limitation of the peace treaty was its failure to take the ethnic principle into consideration, since there were significant masses of ethnic Hungarians left within the boundaries of the new states. On the basis of calculations based on the census of 1910 1,704,851 people were given to Romania, 1,084,343 to Czechoslovakia, 563,597 to Yugoslavia, and 24,807 to Austria. The exact number of foreigners settled or seeking refuge in Hungary after the war, and that of Hungarians emigrating to other countries is not known. The measure of this involuntary migration can, however, be illustrated by the fact that between 1919 and 1923 nearly 200,000 people came to Hungary from the territories annexed to Romania alone.

So the First World War left its mark on the territory and population of Hungary. The huge loss of population was accompanied by the loss of the former multinational character of the country. The number of ethnic Hungarians within the total population became nearly 90 per cent instead of the former 54.5 per cent. The number of Croats, Germans, Serbs, and Slovaks living in Hungary, taken together, was merely 7.4 per cent. The Germans were the most numerous with nearly half a million people. The following table illustrates the distribution of the population by mother tongue according to the results of the cesus of 1920.

Changes in the Number of the Hungarian Population   17

Ethnic Distribution of the Population of Hungary in 1920 by Mother Tongue[6]

Mother tongue	Number of persons	Per cent
Hungarian	7,155,973	89.6
German	550,062	6.9
Slovak	141,877	1.8
Croat	58,931	0.7
Romanian	23,695	0.3
Serb	17,132	0.2
Sorbian and Slovene	6,087	0.1
Gipsy	6,989	0.1
Other	26,123	0.3
Total	7,986,869	100.0

This situation was temporarily changed by the reannexation of Upper Hungary, Subcarpathia, Northern Transylvania, and Southern Hungary between 1938 and 1941. By 1941, the territory of Hungary had grown by 78,680 square kilometers, and the number of the inhabitants by 5,363,331 people, more than half of whom (2,287,032) were ethnic Hungarians.[7]

At the end of the Second World War all this changed once again. Not only did the country lose the reannexed territories, but three more Hungarian villages were annexed to Czechoslovakia. Since then, the territory of the country has been 93,011 square kilometers with roughly ten million inhabitants. The human losses during the last war, and emigration after it, the removal of most of the German nationality, the population exchange with Czechoslovakia, etc. caused a loss of several hundred thousand people. In spite of all these difficulties, the population numbered in 1949 9,204,799, i.e., about 100,000 less than in 1941 calculated for the same territory. The next significant loss was the 200,000 emigrants after the suppression of the 1956 revolution, and the legal and illegal emigration afterwards. Nevertheless, the population was steadily, if not dramatically, growing up to 1980, when it numbered 10,709,463.

National minorities still live in the country, though they are not too numerous. In 1990, when the population of Hungary was 10,374,823, the most numerous nationality was the Gipsy, defined as those whose mother tongue is Roma. Their number was 48,072. They constitute 0.5 per cent of the population. At the same time, the number of those belonging to the Gipsy nationality was 142,683. They were followed by the Germans with 0.4 per cent (37,511 persons), the Croats with 0.2 per cent (17,557 persons), the Slovaks with 0.1 per cent (12,745 persons), and the Romanians with 0.1 per cent (8,780 persons). There were also 3,788 Poles, 2,953 Serbs, 2,627 Slovenes, 1,640 Greeks, 1,370 Bulgarians, 647 Ukrainians and Ruthenes, and 37 Armenians living in the country as Hungarian citizens. The number of people belonging to the rest of the nationalities was 14,570. In 1990 people with the Hungarian as their mother tongue amounted to 98.5 per cent of the total population. Those belonging to the Hungarian nationality were a bit less with 97.8 per cent.[8]

After 1980, however, the low fertility rates and high mortality rates involved that the population started to decrease. Each year Hungary now loses around 30,000 people, the population of a medium-size Hungarian town. This means that in the fifteen years between 1980 and 1995 the country lost 463,786 inhabitants as a natural decrease. The process seems to be irresistable. Without measures encouraging young people to have more children and radical measures stopping and reverting high mortality affecting younger and younger generations, the further decrease of the Hungarian population will become irreversible in the first decades of the next millennium.

Notes

1. Bálint Hóman, Gyula Szekfű, Magyar történet (Hungarian History), (Budapest: Királyi Magyar Egyetemi Nyomda, 1935), vol. 1, 90–181,
2. Gyula László, *A honfoglaló magyar nép élete* (Magyar Life at the Time of the Conquest), (Budapest: Múzsák, 1988), 52–125.
3. Gyula Ortutay, (general ed.), *Magyar Néprajzi Lexikon* (Lexicon of Hungarian Ethnography), vols. 2–3 (Budapest: Akadémiai Kiadó, 1979–80)
4. Gyula Mérei (genereal ed.), *Magyarország története 1790–1848* (The History of Hungary, 1790–1848), (Budapest: Akadémiai Kiadó, 1980), vol. 1, 425–441;
5. Péter Hanák (general ed.), *Magyarország története 1890–1918* (History of Hungary, 1890–1918), (Budapest: Akadémiai Kiadó, 1978), vol. 1, 403–515.
6. György Ránki (general ed.), *Magyarország története 1918–1945* (History of Hungary, 1918–1945), (Budapest: Akadémiai Kiadó, 1976), vol. 8, 414.

7. Ibid.
8. *1990. évi népszámlálás. Demográfiai adatok* (Census of 1990. Demographical Data), (Budapest: Központi Statisztikai Hivatal, 1993), vol. 2, 4–33.

Rudolf Andorka

Demographic Changes and Their Main Characteristics from 1960 to Our Days

To simplify a complex issue, the demographics of Hungary at present are characterized by the following data: in 1994 116,000 children were born (11.3 per thousand inhabitants), and 148,000 people died (14.4 per thousand). Owing to the difference of these figures, the population decreased by 32,000 (3.2 per thousand), which means that the Hungarian population decreases more rapidly than any other country in the world.

However, this statement needs to be qualified. Fertility in Hungary today is not exceptionally low as compared to that of other developed countries, but rather falls in the mid-range. Still, the decrease in the birth rate among the baby-boom generation after World War II occurred much earlier in Hungary than in other countries. Around 1960, fertility was so low that the reproduction of the population seemed to be hopeless in the long run. In most countries of Western Europe, fertility started to decrease in the second half of the 1960s and by the 1970s fell below the level ensuring replacement of the population. Owing to the early decrease of fertility in Hungary, the number of women in their reproductive years is relatively low in the population, and this plays an important part in the low birth rate. It has to be added, however, that unlike some other former socialist countries, there was no immediate drop in the Hungarian birth rate after the recent change of regimes. There was a slight rise in 1990 and 1991, followed by a slow decrease in 1992.

Conditions for mortality in Hungary are, however, exceptionally poor. Mortality rates changed for the worse in the middle of the 1960s, when the male life expectancy at birth decreased by nearly three years, even though it remained basically unchanged with women. By 1990 life expectancy in Hungary, Romania, and the Soviet Union was the worst among the industrial countries. At this time in Austria life expectancy was 7.4 years higher for men, and 5.3 years for women than the rates in Hungary.

In spite of the fact that fertility in Hungary is not exceptionally low and mortality is exceptionally unfavourable, it has to be emphasized that the fundamental cause of the decrease of the population is still the low fertility of women. Should Hungarian mortality be on the Austrian level, the size of the population would still be diminishing, and the difference between the number of births and deaths would be only the half of the present figure.

After this short introduction, let us now examine in detail the recent changes in fertility and mortality in Hungary, and find the possible causes.

Fertility

The simplest index of fertility, i.e., the rate of live births, has fluctuated since 1960, but basically it showed a downward tendency. It touched the bottom first in 1962–64 (12.9–13.1 per thousand), then from 1982 it fell below 13 per thousand again, just to reach the lowest level in 1994 with 11.3 per thousand. From the mid-1960s to the early 1980s it rose somewhat, culminating in 1975 with 18.4 per thousand. The rate of live births is, however, an unreliable index of fertility, since it depends on the age distribution of the population, and it does not express the number of births among a generation of women in their lifetime, but only tabulates the number of births in a given year (see Table 1).

The total fertility rate neutralizes the effects of age distribution, since it is calculated from the age-specific fertility rates for the given year. This index assumes the average number of children that might be born for a generation of women, given the age-specific fertility rate of their generation for that year. The total fertility rate also shows similar fluctuations in Hungary as does the rate of live births, but their amplitude is smaller. It also has the advantage of showing very clearly if fertility in a given year was above or below the level necessary for simple reproduction. If it is around 2.1, fertility is just enough to keep the population stable. Since 1960, actual fertility was thus above the level necessary for simple reproduction only between the years 1974 and 1977. It needs to be mentioned, however, that these higher fertility rates were partly due to the restriction of induced abortion imposed on married women with up to two children. The limitations increased the number of only the first-born and second-born children, so it must have influenced only timing and not long-term reproduction trends. Since otherwise the total rate was mostly 1.8 or 1.9, we can establish that fertility was by 10 to 15 per cent below the level necessary for simple reproduction.

Not even the total fertility rate can express precisely the actual fertility of a generation of women, because it shows the fertility not of an actual generation or cohort, but of a fictive age group giving birth according to the tendency of a given year. The problem arising from the calculations based on a cohort is that we can calculate the fertility only of generations that have completed their fiftieth year of age. Since only a few women give birth after forty, the final fertility rate of women above forty can also be assessed fairly accurately. But nothing can be said about the anticipated fertility rate of the younger generations.

According to Ferenc Kamarás's calculations[1], the total fertility rate for a cohort did not reach the level necessary for simple reproduction already with the generation born between 1931 and 1935. With the generation born between 1941 and 1945 it was only around 1.9, but with the following generations it is not likely to fall further. Should this be really the situation in the future, the deficit calculated in relation to the level necessary for simple reproduction would be only around 10 per cent, i.e., less than the deficit shown by the total fertility rate based on annual data. In other words, if women really have an average of 1.9 children under the present circumstances, fertility must be 10 per cent higher to prevent any decrease in population.

The surveys of fertility and family planning in Hungary give valuable information about the changes in the plans of the families. From an average of 2.33 in 1958, the desired number of children in a family decreased to 1.89 in 1966, then rose to 2.17 in 1974. It decreased slowly again to reach 2.00 in 1987. With respect to the fact that having an average of two children refers only to married women, the desired number of children indicates that the deficit is around 10 per cent of the level necessary for simple reproduction.

What is the cause of the low average of 1.9 births? The direct demographic cause is that most women give birth to no more than two children in their lifetime, and the rate of those having more than two children has been decreasing dramatically in the last two decades (see Table 2). However, this trivial statement means that contrary to popular accounts and debates in the press, the cause of falling fertility rates is not growing childlessness or a conscious desire for parents to have only one child. On the contrary, the number of women choosing to have children has been increasing in the past decades. It is worthwhile emphasizing this factor, since there are countries in Europe where childlessness greatly contributed to the decreasing rate of fertility.

The economic and social causes behind the demographic ones are much more complex to analyze, and no definite answer can be given to them within the framework of this paper. We can only enumerate the possible answers and frame a hypothesis.

Lately, the concept of the "second demographic transition" has come to be widely used in demographic literature. It means that following the baby-boom after World War II, i.e., around 1965 a new trend in falling fertility rates could be observed, and fertility in the developed countries fell below the level necessary for simple reproduction in the long run.[2] The term "second demographic transition" describes in a simplified and concise form the tendencies observed in most developed countries in the last 20–30 years. It offers, however, no explanation for the economic, social, and cultural changes resulting in the falling number of the population.

The fall in fertility rates is not a unique Hungarian phenomenon, but can be observed in almost all developed countries. What is more, in several countries of Europe, the total fertility rate is lower than that in Hungary. In 1994 it was 1.26 in Germany (in former East-Germany it even sank to 0.77 in 1993), 1.22 in Spain, and 1.19 in Italy.[3] It must be added, however, that the total fertility rate per cohort is also higher in these European countries than the rates calculated on the basis of the annual data. The total fertility of the generation born in 1950 in some of these countries reaches or slightly surpasses the level necessary for simple reproduction, but that of the generation born in 1960 is likely to remain below that level everywhere.[4]

The idea of the "second demographic transition" does not offer an explanation for the different tendencies in Hungary, so it is impossible to say why fertility rates fell sooner in Hungary than in other countries and why fertility rates stopped to decrease in the 1970s and after.

Two factors are usually held to be the cause of the second demographic transition: first, the availability of modern contraceptives; second, the change of attitudes, values, ideals, and goals of life. Contraceptives do not account for the fall of fertility in Hungary by themselves, for the decrease took place before they reached the public at large. When induced abortion was made nearly legitimate in 1956 (it was allowed to be performed at the request of the women in the first three months of the pregnancy), it must have contributed to the decrease of fertility rates, even though it was only one of the means of limiting the desired number of children. The valid and important question is why the couples wanted to have fewer children than before.

This leads us to the problem of attitudes, norms, and values. The simplest explanation based on these factors declares that the number of children decreases because the couples decide to have fewer children, for reasons of their own.[5] This can be proved clearly by empirical data. The ideal number of children in surveys on family planning in developed countries where fertility has fallen below the level necessary for simple reproduction is around two, its average is somewhat above 2.0. A survey from 1977 done in Hungary in the framework of the World Fertility Survey shows that the Hungarian families of today hold an average of 2.46 children desirable, none of them vote for childlessness, only 0.9 per cent votes for one child only, 54.7 per cent for two children, 42.3 per cent for three, 1.8 per cent for four, and 0.3 per cent for five children or more.[6] The interviewed persons considered, however, fewer children desirable for their own families. Since there were no similar surveys prior to the 1960s, we do not know anything about the ideal or desired number of children in Hungary in the first half of the twentieth century and in the nineteenth century prior to the demographic transition, but one can safely conclude that it must have been more than two or three.

It remains to be seen why this ideal number began to decrease. The most commonplace theoretical explanation would be the growing employment of women outside the household. Employed and wage earning women really have fewer children in Hungary today than homemakers. It seems to be obvious that working outside the household cannot be reconciled with bringing up several children. Another explanation that goes somewhat deeper is that a change has occured in the mentality and aspirations of women. Work has become increasingly more important for them than having several children. This hypothesis cannot be proved by empirical data as yet, but the available data show that most women want to have two children and an occupation at the same time.

Even deeper go the hypotheses that indicate spreading individualism and anomia in modern society. Its representatives maintain that human relations based on communities (*Gemeinschaft*) are giving way to ones embedded in society (*Gesellschaft*), whereby the importance of the family is diminishing or heads toward a crisis.[7] In consequence of all this, people tend to find individual success and short-term joy more important, and the happiness of others (their partners, children or parents) less and less significant. Couples are more and more dominated by short-term benefits, and bringing up children is pushed to the background, since it is, by definition, a long-term constraint. Although several scholars speak of this tendency in the developed societies today, empirical data are not available to prove it. When trying to use this hypothesis to explain the fall in the fertility rates, one cannot help asking the question why modern adults should have any children at all. While accepting the tendency of individualization as one of the causes or at least as the social background of the decrease of fertility, one has to accept the existence of a counter-factor, which causes the majority of modern people to have children. The question is, how many children they want to have.

The costs of bringing up children are to be considered at this point. When these costs rise faster than the income of a family, the parents undertake to have fewer children than they planned. In other words, those who have one child mostly do not want to have a second one, and ever fewer families with two children undertake to have a third.

I do not agree with the highly simplified economic approach to fertility put forth by G. Becker.[8] Parents do not decide the number of their future children as they decide buying durable consumer goods, as Becker presumes, which means that they do not consider benefits and costs only. Should this be the case in modern society, there would hardly be any children at all, since the costs of bringing up children are much higher than the material benefits for the parents. If we take also the non-material benefits into consideration, we enter the realm of non-measurable phenomena and return to attitudes and values.

The costs of bringing up children are undoubtedly rising in modern society, partly owing to longer education. Young people in their twenties increasingly need their parents' assistance even after completing their school years.[9] Conversely, in present day society children are of hardly any financial benefit for parents. When they become wage-earners themselves, they become financially independent from their parents. Old age pensions make their contribution to the well-being of their elderly parents less important.

The very early decrease in fertility in Hungary in the second half of the 1950s must have been due to the rise of the costs of bringing up children, and to the longer years at school, in addition to the unfavourable changes in the income of the families and the general economic and political uncertainty.

Still more important is the short-term relationship between the costs of raising children, the income of the families, and the number of children they have. In my opinion, these factors have played an important role in the changes of fertility since 1960. These changes are naturally slight, since the total fertility rate calculated on the basis of annual data varied between 1.69 and 2.38, while the fluctuation of the total fertility rate per cohort was still smaller. However, even these slight changes influence the demographic situation of the country to a considerable degree, since a value of 2.1 means that the population is able to reproduce itself in the long run, while a value of 1.7 means that the generation of the children is by 19 per cent smaller in absolute numbers than that of the parents, i.e., each generation is by 19 per cent smaller than the other.

The slight rise in the total fertility rate and the desired number of children after the nadir in the middle of the 1960s were presumably due to the significant extention of family benefits in the second half of the decade. The value of family allowance was gradually rising both nominally and comparatively with the wages, and a so-called child-care allowance was also introduced, making it possible for working mothers to stay at home with the children for three years on a social allowance approximately equal with the minimal wages, and to return to their former job when the child became three. The slight decrease in the total fertility rate and the desired number of children from the late 1970s were, however, due to the fact that the value of the child-care allowance, and that of the family allowance started to decrease as compared to the wages. In the 1990s even their real value started to decrease.

The changes of income must have had an influence on the above processes in fertility, too, but it cannot be proved empirically as yet. From the second half of the 1960s to the mid-1970s the average level of income was rising more rapidly than in the previous years. The real wage index was, however, falling from 1978 onward and in the 1980s the growth of the incomes slowed considerably. In the 1990s, the per capita income has decreased by about 12 per cent. A relatively higher fertility index can be observed in the years of quicker growth, while the decrease of incomes coincides with the decrease of fertility.

The impact of family benefits on fertility is even more important than these short-term fluctuations. Their introduction and increase from the second half of the 1960s stopped the falling tendency of fertility. In the 1970s and 1980s all structural changes of society contributed to the decrease of fertility. The percentage of people with higher educational levels, intellectual occupations, and the number of town-dwellers was steadily rising, and people in these categories usually have fewer children. However, in these decades nearly all social categories tended to have more children, which counterbalanced the fact that an ever decreasing number of women belonged to the categories usually with several children. The educational level of the mothers clearly reflects this trend. From 1970 onward the fertility of married women on each educational level increased (see Table 3). One can draw the conclusion that in absence of family benefits, the level of fertility would be much lower than it is now.

Mortality

The reasons why mortality rates increased are well known. Between 1960 and 1992, 25 per cent of the highest male mortality in the age group 50–54 can be attributed to coronary diseases, another 25 per cent to cirrhosis of the liver, 14 per cent to lung cancer, 8 per cent to "clay-pipe cancer", cancers of the mouth cavity, the pharynx, and the oesophagus, 8 per cent to cerebral diseases, and 4 per cent to suicide.[10] Alcoholism and smoking, along with hypertension must have contributed to all these factors. It is difficult to say how these phenomena are related to one another. Hypertension is presumably caused by an unhealthy diet, the lack of exercise, smoking, and excessive drinking. In the background of heavy drinking and smoking, as well as of the growing suicide rate there is a stressful way of life. Moreover, the lack of proper medical care cannot be excluded, although definitive evidence is lacking.

To sum up, the main causes of increasing mortality rates in present-day Hungary are directly related to the crises in the society at large. This general crisis is most probably connected to the political history of the past decades, namely to the totalitarian (later authoritarian) character of the regime that made many lives unnecessarily tense and impaired the health of its members, and made proper health care impossible.

Demographic, economic, and social consequences of the demographic situation

All consequences of the demographic situation of the country manifest themselves in the middle and even more in the long run. This remoteness of cause and effect make it difficult to ameliorate demographic problems. The possible negative consequences to be expected after several decades are generally not reckoned with neither by the constituents (many of whom will no longer be living when the problems arise), nor the governments and politicians depending on the former.

When dealing with the consequences of the demographic processes today, one has to concentrate on such long-term effects, although the consequences of the situation that came about in the 1960s seem to be ripening now.

There is of course no guarantee that the demographic situation will be altered in the near future. Only mortality can be predicted more accurately, and a slow improvement is to be expected. It can, however, not be taken for certain, either, since stagnation on the present very unfavourable level cannot be excluded. Things can even turn for the worse, as observed in some former socialist countries. As regards fertility, a significant rise can most probably be excluded. The number of births is not likely to return to the level of the early twentieth century that was much above the level necessary for simple reproduction. The highest predictable fertility level may lead to simple reproduction, which is a 10 to 15 per cent increase in the fertility rate. A further significant decrease of fertility cannot be excluded, either, considering the possible fall of the incomes and the reduction of state support beginning with 1995. When outlining a few possible consequences I start out from the hypothesis that fertility remains unchanged.

The primary demographic consequence of this will be the decrease of the present population. (The actual number of the population can vary because of possible international migrations to be mentioned later.) It is possible to calculate the changes of the population in the following decades[11] and to see when the population diminishes altogether, as Bourgeois-Pichat[12] demostrated in his studies of world population. So what Chesnais[13] calls, in Raymond Aron's words, the suicide of the Europeans might be predicted in the light of the present tendencies of low birth rates. This is naturally not what will happen, but merely a supposition in case the present tendencies remain unchanged.

The secondary demographic consequence is the aging of the population, i.e., the growth of the rate of those above retirement age, and the decrease of the number of children. In 1990, the ratio of those above sixty was 18.9 per cent, which meant that there were 35.6 people of age sixty and over per one hundred between twenty and fifty-nine. Both proportions belong to the highest in the world, and if our mortality rate were not so unfavourable, they would be higher still. If the present fertility rate remains unchanged and mortality hopefully improves, the ratio of the elderly will be still higher in the future.

The aging of the population has and will have several far-reaching effects on society and economy, most of which have not been proved unequivocally yet. One of them is, however, clear enough: the higher the rate of the old people in a population, the larger the deficit of the superannuation funds, unless the present system of financing pensions is changed.

Today, in most developed countries, Hungary included, there is a "pay-as-you-go" system of pensioning, which means that a pension is paid from the contributions of the earners. In the beginning, such a system is very favourable, since those who pay, are due to get pension only decades later, and those who get pension right then, have not paid for it earlier and get, so to speak, a present. But as those paying their contributions will be entitled to get pension in an ever growing number, it is quite natural for them to feel entitled to get a fair pension, and feel themselves deceived if they do not get the appropriate sum. Given a stationary population (i.e., when the number of births is high enough to ensure generations of the same size following one another), and under the present conditions of mortality in Hungary today, a man retiring at the age of sixty should have paid thirty per cent of his income during forty years of work to be able to get a pension equalling eighty per cent of his real income for the rest of his life, which is expected to be fifteen years. The present contribution to retirement pension funds paid both by the employers and the employees, although very high as compared to other countries, and should cover the pensions and the costs of health care, is about one third of the total expenditure on wages. It becomes clear from the above that the present superannuation system cannot be maintained in the long run. Either retirement age should be raised considerably or pensions should be cut, or contributions should be raised.

The problems of the superannuation system can be eased by the rise of the real wages, for in that case the contributions also grow. But the situation is aggravated if fertility is low and does not reach the level necessary for simple reproduction, for in that case the generation entering the labor market is smaller than the one leaving it, and a decreasing number of employed is going to contribute less. Low fertility rates thus contribute to the crisis of the superannuation system.

The population will naturally not decrease if the number of immigrants from other countries is equal to the difference between the number of births and deaths. The above mentioned problems due to aging would also disappear or at least be mitigated if young adults immigrated into the country. It seems to be nearly imperative for Hungary to receive as many immigrants as the deficit of the Hungarian population will be. These immigrants are economically beneficial for the aging population. However, these foreigners, especially if they are very numerous and belong to a very different culture, may cause tension in Hungarian society. So it would be much better if the difference between births and deaths were only minimal.[14]

Conclusions

To sum up, Hungarian fertility is low, and it would be desirable for it to be higher and to reach the level necessary for maintaining the present population, which could be facilitated by increasing family benefits. There is also another reason for increasing family benefits, namely, the living standard of families with children, especially of those with several children, has gradually been falling in the past few decades, and especially since the recent change of regimes, as compared to the average of the population and even to the living standard of the elderly people. So when discussing changes in the family benefits, one should remember the words of S. H. Preston, who urged American demographers and the public to decide if they wished to care only for their own future, in which case the impoverishment of the families with children would be really irrelevant, or wished to consider also the future of the commonwealth, in which case family benefits would have to be raised.[15]

Table 1
Some Primary Demographic Indexes between 1960 and 1994

Year	Live Births	Deaths	Natural Increase, Decrease	Total Fertility Rate	Life expectancy at Birth in Years	
	per thousand inhabitants				Male	Female
1960	14.7	10.2	4.5	2.02	65.89	70.10
1961	14.0	9.6	4.4	1.94	66.71	71.09
1962	12.9	10.8	2.1	1.79	65.60	70.02
1963	13.1	9.9	3.2	1.82	66.61	71.20
1964	13.1	10.0	3.1	1.81	67.01	71.78
1965	13.1	10.7	2.4	1.81	66.71	71.54
1966	13.6	10.0	3.6	1.88	67.53	72.23
1967	14.6	10.7	3.9	2.01	66.92	72.04
1968	15.1	11.2	3.9	2.06	66.69	71.94
1969	15.0	11.4	3.6	2.04	66.68	72.00
1970	14.7	11.6	3.1	1.96	66.31	72.08
1971	14.5	11.9	2.6	1.91	66.11	72.04
1972	14.7	11.4	3.3	1.93	66.85	72.57
1973	15.0	11.8	3.2	1.95	66.65	72.49
1974	17.8	12.0	5.8	2.32	66.52	72.38
1975	18.4	12.4	6.0	2.38	66.29	72.42
1976	17.5	12.5	5.0	2.26	66.60	72.50
1977	16.7	12.4	4.3	2.17	66.67	72.99
1978	15.8	13.1	2.7	2.08	66.08	72.74
1979	15.0	12.8	2.2	2.02	66.12	73.03
1980	13.9	13.6	0.3	1.92	65.45	72.70
1981	13.4	13.5	-0.2	1.88	65.46	72.86
1982	12.5	13.5	-1.0	1.78	65.63	73.18
1983	11.9	13.9	-2.0	1.73	65.08	72.99
1984	11.8	13.8	-2.0	1.70	65.05	73.16
1985	12.3	14.0	-1.6	1.83	65.09	73.07
1986	12.2	14.0	-1.8	1.83	65.30	73.21
1987	12.0	13.6	-1.6	1.81	65.67	73.74
1988	11.9	13.4	-1.5	1.79	66.16	74.03
1989	11.9	13.9	-2.1	1.78	65.44	73.79
1990	12.1	14.1	-1.9	1.84	65.13	73.71
1991	12.3	14.0	-1.7	1.86	65.02	73.83
1992	11.8	14.4	-2.6	1.77	64.55	73.73
1993	11.4	14.6	-3.2	1.69	64.53	73.81
1994	11.3	14.4	-3.1	1.64	64.84	74.23

Table 2
Average Number and Distribution of Children of Women between 45 and 49 at the Time of the Censuses Between 1960 and 1990

Year of Census	Live Births per Hundred Women	0	1	2	3	4	5	6-9	10-X	Total
				Live Birth (percentage)						
1960	264	13.5	20.2	25.0	16.0	9.4	5.5	8.4	1.9	100.0
1970	227	10.3	22.8	32.8	17.4	8.1	3.9	4.2	0.5	100.0
1980	202	7.4	27.0	41.7	14.5	5.0	1.9	2.1	0.4	100.0
1990	196	5.0	25.0	50.8	13.4	3.3	1.1	1.2	0.2	100.0

Table 3
Number of Children per Hundred Married Women in Selected Age Groups, According to Educational Level, 1960–1990

Age	Education	Number of Live Births per Hundred Married Women			
		1960	1970	1980	1990
25-29	Primary school	148	147	178	263
	Secondary school	111	108	137	145
	Higher education	94	93	112	120
	Total	161	145	158	161
30-34	Primary school	174	174	200	299
	Secondary school	150	139	165	176
	Higher education	135	133	159	170
	Total	205	184	188	193
35-39	Primary school	183	187	195	214
	Secondary school	177	147	162	183
	Higher education	160	152	166	180
	Total	233	205	194	200

Notes

1. Ferenc Kamarás, *"Longitudinális vizsgálatok szerepe a népesedési folyamatok elemzésében. A gazdaság intenzív fejlődése és a statisztika c. konferencia előadásai"* (Role of Longitudinal Surveys in the Analysis of Demographic Processes. Lectures of the conference "The Intensive Development of Economy and Statistical Research") (Budapest: KSH, 1984), 97–104.
2. D. J. Van de Kaa, "Europe's Second Demographic Transition," *Population Bulletin* 42, no. 1 (1987): 59.
3. C. Guibert-Lantoine, A. Monnier, "La conjoncture démographique: L'Europe et les pays développés d'Outre-Mer," *Population* 50, no. 4–5 (1995): 1185–1210.
4. L. Roussel, "Fertility and Family", in: *European Population Conference Proceedings*, vol. 1 (New York: United Nations, 1994), 35–110.
5. J. Blake represented this theoretical standpoint in such a very simple form in his article criticizing the economic approach to fertility under the title "Are Babies Consumer Durables?", *Population Studies* 22, no. 1 (1968): 5–25.
6. Világ termékenység vizsgálat: Magyarország adatai (World Fertility Survey: Data for Hungary) (Budapest: KSH, 1982), 254.
7. H. J. Hoffmann-Novotny, "The Future of the Family", in: *European Population Conference 1987, Plenaries* (Helsinki: Central Statistical Office of Finland, 1987), 113–200.
8. G. Becker, "An Economic Analysis of Fertility", in: *National Bureau of Economic Research: Demographic and Economic Change in Developed Countries* (Princeton: Princeton University Press), 209–240.
9. The Sociological Institute of the Bamberg University and the Sociological Department of the Budapest University of Economics examined the parental support given to young people between 18 and 29 in Eastern and Western Germany, and Hungary. We have found that this support is very significant in all of these societies. See L. A. Vaskovics, R. Andorka, Zs. Spéder, Gy. Bárdossy, G. Pickel, *Intergenerative Solidaritätsbeziehungen in der Familie: Junge Erwachsene und ihre Eltern.* (Unpublished).
10. Péter Józan, *A halálozási viszonyok alakulása Magyarországon 1980–1992* (Changes in Mortality in Hungary, 1980–1992) (Budapest: KSH, 1994), 124.
11. See the article of László Hablicsek in this volume.
12. J. Bourgeois-Pichat, "Du XXe au XXIc siecle: L'Europe et sa population apres l'an 2000", *Population* 43, no. 1 (1988): 9–44. According to his calculations, if the fertility rate sinks to the level of 1.2 children per mother in Western Germany at the time of the writing of the article first in the other industrial countries, then all over the world, the population of the former

will die out around 2250, and that of the whole world around 2400. He naturally did not say by this that this is sure to happen.
13. J.-C. Chesnais, *Le crépuscule de l'Occident* (Paris: Robert Laffont, 1995), 10, 17.
14. R. Lesthaeghe, "Are Immigrants Substitutes for Births?", Paper presented at the Symposium on Population Change and European Society (Florence, 7–10 December, 1988).
15. S. H. Preston, "Children and the Elderly: Divergent Paths for America's Dependents", *Scientific American* 251, no. 6 (1984): 44–49.

Magdolna Csernák

Marriage and Divorce in Hungary: Demographic Aspects of Changes

Introduction

Recent demographic trends for a considerable number of European countries suggest a remarkable change in family life. During the last few decades European fertility has declined to levels below replacement of present population and no sign of a sizeable and lasting upturn has yet been perceived. This unprecedented phase of population history, called by some demographers a "second demographic transition", has been accompanied by major alterations in many areas of family life such as the formation and dissolution of marriage.

From the end of the 1960s onwards the marital behaviour across the continent began to change rapidly. Marriage rates have fallen dramatically and more and more men and women postpone first marriage or reject it and many of them choose living together with a partner without marriage. The proportion of marriages ending in divorce has also radically increased. Simultaneously with these phenomena a downward trend has started in remarriages. The weakening of the traditional family is marked by not only a declining propensity to marry and more frequent divorce but also increasing childbearing without wedlock. One-parent families and living alone have also gained considerable importance.

These developments are closely interrelated and they can be explained by increasing plurality of lifestyle decisions.[1] The explanations, however, primarily stand for Northern and Western Europe where there has been a massive transformation in values and attitudes in the last thirty years. Premarital sexual relationships have become the norm, marriage has become less popular, the acceptance of "sleeping together" has evolved into tolerance of "living together", often as a prelude to marriage.

In contrast to these developments, marriage has remained popular in Eastern Europe and the traditional marital behaviour has been relatively unchanged. Until recently, the countries of the Eastern block have formed a relatively homogenous group, although there is some evidence that the recent political and social changes have also effects on marital behaviour. In this respect Hungary seems to be a forerunner in East Central Europe because changes in marriage have started earlier than in the neighbouring countries and an increasing convergence might be expected with the predominant European pattern.

From point of view of population development and family life, the following demographic features of nuptiality (formation and dissolution of marriage) require more attention:

- Until recently the Hungarian population has been characterized by an early and almost universal marriage. However, during the last decade a radical turn has been taken for the opposite direction. The downward trend can be observed both in first marriages and remarriages.
- The frequency of divorce began to increase immediately after the Second World War, coinciding with major changes in the divorce law and in the economic and political system. Nowadays in Hungary divorce is a usual and accepted way of ending marriage.
- In Hungary new life styles are also spreading. Although, many young never married men and women choose living together without wedlock it is still more frequent for ever married (divorced and widowed) population.

The aim of this paper is to provide a picture of recent demographic changes in marriage and divorce and contribute to a better understanding of the role of nuptiality in the family relationships in Hungary. For this reason when describing the current trends special attention is given to some background factors which may, to a certain extent, explain the peculiarities of the Hungarian situation.

Sources of data

The data presented here are primarily derived from vital statistics registered by the Hungarian Central Statistical Office (HCSO). In addition, information was also used from calculations and estimations carried out in the HCSO Demographic Research Institute based on secondary analyses for statistical data.

The limited space for analysis means a restriction which allows only a brief overview on main trends of nuptiality. As a consequence some important aspects of nuptiality has been neglected such as the gender dimensions, the "marriage squeeze" concept or the implications of nuptiality for other areas of family life.

Historical pattern of marriage in Hungary

The turn of the twentieth century Europe was characterized by two distinct marriage patterns. The imaginary line from St. Petersburg/Leningrad to Trieste that divided Europe into two parts was drawn by the demographer John Hajnal in his stimulating work published in 1965.[2] Northwest of this line households were small and nuclear and the family formation was connected to the rule that the newly-wed couples had to be able to establish an independent household. The requirement of economic independence practically restricted them from marrying early and couples tended to marry at relatively advanced ages while a high proportion of population never married at all.

In contrast with this so-called "European" pattern of marriage, in most part of Eastern Europe, where the low standards of farming and other economic and social factors stimulated the young married couples settle in the farmstead of the parents and join the family production: The economic strategy of these families required a much earlier and near-universal marriage. Since this marriage pattern was wide-spread throughout the world, independent of the geographical situation, social and economic systems, Hajnal called it a "non-European" pattern. In the European context, however, the distinction "Eastern European" has come into use.

Hungary being geographically in the East Central part of Europe followed the Eastern European marriage pattern at the beginning of the century. Most men married at age of 20–24 years and most women at about 20, and only about 4–5 per cent of them remained unmarried throughout their lifetime.

During the first half of the century the Hungarian population abandoned this scheme: both the mean age at first marriages and the proportions of ultimately single population increased. The changes in timing and intensity of first marriages started in the first decade of the century. The shift into the direction of later and less-universal marriage became stronger especially after the First World War, due to the postwar economic difficulties and the great economic crisis of the early thirties. Between the two world wars, the mean age at first marriage increased to 28 years for men and 24 years for women. The proportion of the ultimately single population also increased. In the cohorts married between the two world wars about 7–8 per cent of women remained single throughout their childbearing years.

The differences between the two marriage systems decreased by the 1960s: barriers to young and universal marriages in the North and West disappeared, as the industrialisation and urbanisation processes and the practice of birth control spread across Western Europe. This trend towards a "young and universal" marriage continued in the post war period, reaching its zenith during the early 1960s, which was described as the "golden age of marriage" in Western Europe.[3]

Marriages in Hungary after the Second World War

In measuring the annual flow of marriages the simplest index is the crude rate of marriage (marriages per 1000 of mid-year population). Taking the Second World War as starting point, the frequency of marriages increased after a short and not too deep decline during the years of the war. The marriage "boom" surpassed that of most European countries and the marriage rates remained at high levels (9-10 marriages per 1000 mid-year population) also after the compensation of losses in marriages due to the War. A slight downward trend appeared only at the end of the 1950s which was followed by another upward tendency with a local peak at the second half of the 1960s. After a strong wave of marriage in the mid-1970s, which was partly due to the large cohorts born in the 1950s entering marriageable ages, a steady decline has started in the intensity of marriage. (Table 1)

For quite a few years Hungary has lost its special place in marriage in Europe. The crude rate of marriage was 5.3 per 1000 mid-year population in 1994, and the decreasing trend seems not to have stopped. During the last decades the marriage rates took a similar downward direction all over Europe, and Hungary with its moderate level of marriage conforms to this.

Although crude rates are useful as gross measures of relative frequency of marriages, they cannot explain all the variations of marriage trends. Some distorting factors contributing to variations in crude rates, for example, are the proportion of the population in marriageable age, the proportion of previously married, the area's economic situation, provisions for dissolution of marriage contracts, and customs pertaining to remarriage. For this reason it is necessary to pay attention to some more specific indicators of marital behaviour.

Marriages by marital status

In Hungary the overwhelming majority of population marrying is single and contracts marriage for the first time. Thus the pattern of first marriages determines essentially the main characteristics of marital behaviour. However, remarriage is also significant in living arrangements and it has always been an accepted institution in Hungary. As a consequence, the heterogeneous composition of marriages by marital status is one of the characteristic features of Hungarian marital behaviour.

At present one quarter of marrying parties had been already married previously. The proportions of ever married (divorced, widowed) population among people marrying have been almost unchanged during the last decades. This phenomenon has mainly been in connection with the increasing tendency in divorce. At the same time the propensity to remarry in the widowed and divorced population has continuously declined.

In the past, remarriages of the widowed population dominated due to the unfavourable mortality in Hungary. After the Second World War, however, in both absolute and relative sense the remarriages after divorce became significant. At the 1960s three quarters of divorced men and about two third of divorced women married again within 20 years after the divorce. Roughly up to the 1980s, remarriage was a considerable factor in high proportions of married population in Hungary. Since the 1980s cohabitation as new living arrangement has spread in the divorced population and, as a consequence remarriage rates have dropped to very low levels. If the remarriage rates of divorced population were to continue at the level of 1994 less than 40 per cent of them would remarry in the future.

In Hungary widowed people rarely remarry nowadays. Only about 7–8 percent of men and 2–3 per cent of women remarry after the death of husband/wife.

In this paper we shall restrict ourselves to highlighting selected data on first marriages. Two broad aspects of first marriages will be discussed here: the intensity of first marriage, that is, what proportions of single (never married) population get married until the age 50; and the timing of first marriage, that is, at what age people get married.

To measure the intensity and timing (or tempo) of first marriage there are some classical methods. The most popular indices are the total first marriage rate and the mean age at first marriage. (For definition of the terms see Table 1).

Proportions first marrying

Total first marriage rates can, under special assumptions, be interpreted as the life-time number of first marriages that a hypothetical (synthetic) cohort would experience if the age-specific first marriage rates remained the same as those found during the given calendar period. Total period first marriage rates have been available for Hungary from the beginning of the 1950s. These rates are appropriate to illustrate the effects of the high frequency of marriage on the proportions first marrying in Hungary. (Table 1)

The high level of first marriages in Hungary can be explained by some side effects of radical social and political changes after the Second World War. For example, the great social mobility which broke down some barriers between the social strata, and the huge influx of women into labour force have increased the propensity to marry. It should be also noted that from ideological purposes, among others, in order to reduce the parental authority a radical reduction has been introduced in the legal minimum age at marriage in 1952. The minimum age for contracting marriage was reduced from 24 years to 18. In addition, legal permission was granted to marry under this age for particular reasons.

During the first half of the 1960s an opposite tendency prevailed. Tensions connected with the collectivization of agriculture and with a smaller "disequilibrium" in the size of cohorts caused a trough in the trend of the total first marriage rates. However, by the second half of the decade the rates increased again in connection with some improvements in living standards and with the introduction of child care allowance in 1967, which increased the material advantages in marrying and having a child.

Marriage and Divorce in Hungary 43

Table 1
Selected indicators of marriages
1948–1994

Years, average of years	Number of marriages	Marriages per 1000 mid-year population	Total first marriage rate (TFMR)* per 100		Mean age at first marriage** (in years)	
			men	women	men	women
1948–1949	102,765	11.2	..	..	26.4	22.8
1950–1959	98,235	10.2	112.8	110.6	25.7	22.3
1960–1969	89,523	8.8	98.5	96.5	24.7	21.6
1970–1979	97,097	9.2	91.7	96.1	23.7	21.0
1980–1989	72,854	6.9	76.5	82.9	24.3	21.4
1990	66,405	6.4	77.1	77.2	24.2	21.5
1991	61,198	5.9	70.2	70.4	24.2	21.5
1992	57,005	5.5	65.1	64.5	24.3	21.6
1993	54,099	5.3	60.3	57.8	24.4	21.7
1994	54,114	5.3	58.9	56.4	24.7	22.0

Source: Volumes of Demographic Yearbook, Central Statistical Office, Hungary

.. Data are not available.

* **Total first marriage rate**: sum of age-specific first marriage rates per one (or 10, 100 etc.) head of the respective population between ages 15 and 50. More than one first marriage per person (see the years 1950–1959) means a fluctuation due to a short-term increase in the number of marriages in the given period.

** **Mean age at first marriage** is obtained by the weighted average of ages at marriage, weights being the number of first marriages at each age.

In the 1970s, a reversal to the old customs was supported also by modification in family law. In 1974, the legal minimum age of marriage was reduced to age 16 for women. The measure was supposed to follow traditions of Hungarian society. It also intended to prevent births out of wedlock. This modification has coincided with the introduction of new population political measures which aimed to increase fertility and indirectly contributed to a fluctuation in marriage in the mid-1970s. As a consequence, paralleling the radical decline which started in first marriages all over Europe, in Hungary on the yearly average, 15 per cent of women married by the age 18 and 40 per cent of them married by the age 20. The differences between Western Europe and Hungary become wider in the 1970s.

The upward trend of the mid-1970s lasted, however, only for three years as more signs of changes appeared in marriage behaviour. The downward trend started after 1975 and it was strengthened by a further change in family law at the second half of the 1980s. The last modification has increased the minimum age for marriage and abolished the possibility of marrying under age 18 without legal permission.

Analyzing the age-specific first marriage rates there are some signs of a decline in early marriages in Hungary. Especially the proportion of women marrying under the age 20 has fallen. At the beginning of the 1990s, the proportion of women marrying under 18 was 4 per cent and only 26 per cent of them married before attaining the age of 20. If the age-specific first marriage rates of 1994 were to continue more than 40 per cent of men and women would not marry at all until the age 50.

Current trends of first marriages in Hungary show many similarities to the "quiet" revolution that started more than two decades ago in Western Europe. The rapidity of changes is especially striking. Up to 1993 the total first marriage rate already reached the same level characterizing the western countries at the second half of the 1980s. It means that the radical decline has occurred within a very short interval. Altogether, the frequency of marriage indicates similar tendencies to that of western part of Europe, and Hungarian marriage behaviour seems to fit into the universal European marriage pattern.[4] It is probable, however, that we can speak rather about statistical similarities. There are some questions in connection with the similarities of values behind recent trends which cannot be answered without further research. One of the most significant features is that in Hungary the more significant changes in behaviour towards marriage have appeared in parallel with the deep economic recession of the 1990s. The answer to the question "to marry or not to marry" does not mean free choices between life-styles for many young people.

Undoubtedly, however, cohabitation has increasing popularity among the young and it plays a significant role in the declining propensity to marry. Recently remarkable changes occurred in the proportions of those living in cohabitation among the youngest population. Comparing the number of cohabiting women to the total number of women living in union (together married and non-married relationships) in 1984, 12 per cent of women under the age of 20 lived in cohabitation. In 1990, already every fourth had a cohabiting partner among the corresponding women living in union.

Mean age at first marriage

An important summary index of marriage patterns is the mean age at which people marry. The trend with respect to age at first marriage in Hungary between 1950 and 1993 is illustrated in Table 1. Clearly the trend has been overwhelmingly downward, towards younger ages of marriage for both men and women until the 1970s. During the second half of the 1970s the trend reversed. In 1994, the mean age at first marriage was 25 years for men and 22 for women. However, it is still much lower than in western part of Europe, where the mean age at first marriage reached 29 years for men and 27 years for women in several countries at the beginning of the 1990s.

It is questionable why the period mean age at first marriage has increased so slowly and why it has remained at a rather low level up to now. One explanation for this is that the postponement of marriage can be experienced almost in every cohort not only those who are at the young marriageable ages.

Cohort perspective of first marriage

On the basis of period data it is risky to answer the question whether the declining tendency of marriages is a transitional phenomenon stemming from the temporary postponement of marriage or whether it indicates a substantial long-term decline also in the frequency of marriage. Cohort analysis of first marriages offers supplement information to these questions.

The cohort data of Table 2 for Hungarian women indicate the same tendencies in both the ultimate proportion of first marriages and the mean age at first marriage as it can be experienced based on period data. However, the final level of cohort nuptiality seems a relatively more stable value and the changing cohort behaviour modifies only slowly the proportions remaining single at higher ages. In other words, the period fluctuations, the peaks and troughs in the trend of first marriage mainly result from transitional changes in the timing of first marriage.[5]

Table 2

Mean age at first marriage and ultimate proportions first marrying per 100 single women by birth cohorts*

Year of birth	Mean age at first marriage (in years)	Ultimate proportions first marrying
1900–1909	23.15	92.58
1910–1919	23.43	93.80
1920–1929	22.82	95.37
1930–1939	21.68	96.06
1940–1949	21.37	95.50
1950–1959	21.30	94.72
1960–1969	21.65	90.04
1970–1974	23.38	81.48

* Figures based on cohort nuptiality tables for Hungarian women. The estimated values of residual nuptiality for the cohorts born between 1945 and 1974 have been calculated by using period age-specific first marriage probabili-ties for 1993/1994 (i.e. based on assumption of constant nuptiality after 1993/1994)

According to this estimate, the youngest cohorts may abandon the traditional patterns of first marriage. In the cohorts born in the 1960s a radical fall in marriages may be expected. The ultimate proportion of 80 per cent for ever marrying would be historically unprecedented in Hungary. Further alterations may occur in the age distribution of first marriages. From the cohorts born in the second half of the 1960s onwards the mean age at first marriage may reach 24 years. This figure is higher in cohort perspective than in period one. It should be noted, however, that it is surprisingly similar to that of the women born at the beginning of the century.

Dissolution of marriage

A marriage may come to an end either by the death of a spouse, or by a legal procedure. The principle method of legal termination of a marriage is by divorce. Most existing marriages are dissolved by the death of one of the partners, usually by the husband's death. In Hungary married population is strongly affected also by the unfavourable general mortality. Due to the high level of divorces and dissolutions by death the balance of nuptiality has been negative since the beginning of the 1980s in Hungary. Thus the number of dissolved marriages by divorce and death exceeds the number of contracted marriages.

Taking into consideration the remarkable role of divorce in the marital relationships and the growing diversity of life-styles in the following section a special attention will be given to divorce.

Divorces after the Second World War

Legal divorce has been recognized in Hungary since the end of the 19th century. During the first half of the century the legal termination of marriage was based on the faults of a spouse. After the Second World War this system was abolished and a new liberal legislation based on no-fault criteria was introduced. Under the Family Act of 1952 a marriage could be legally dissolved if it had irretrievably broken down. This reform shifted the burden of deciding the "serious and well-founded reasons" onto wives and husbands themselves.

In the light of judicial practice divorce law was further liberalized in 1974. Since then divorce by mutual consent of the spouses has been permitted. A relevant factor in deciding this reform was that a further deterioration in the relationships between the spouses might be harmful to the development of any child of the family.

The unchanged legal frames and the wide recognition of liberty to get a divorce offer an opportunity to draw a rather clear picture on changing behaviour toward marriage in Hungary. As the Hungarian law promotes the dissolution of broken marriages, it is very rare that marriages which have lost their stability would be maintained even formally.

By international standards the number and rates of divorce in Hungary were already high in the past even before the First World War. Hungary's dominant position in this respect continued until the 1970s in spite of the fact that the divorce rates in other European countries, first of all in Eastern Europe, have also increased. From the 1970s onwards marriages in Western Europe, too, became noticeably less stable and laws relating to divorce were liberalized. The number of divorces has increased considerably throughout Europe and Hungary takes an average position compared to other European countries at present.

One of the most frequent indicators used in demographic analyses is the total divorce rate. (Table 3) For period it can be interpreted as the number of divorces that would occur if the duration-specific divorce rates of the given period were to remain constant. The total divorce rate is a standardized index which eliminates distortions resulting from the different sizes of marriage cohorts following each other. At the same time it slightly underestimates the level of divorce because the population of marriage cohorts is reduced also by mortality not only by divorce. Another defect of this measure is its sensitivity to timing, but in spite of this the total divorce rate may be preferred for analyzing changes in divorce.

Since the beginning of the 1950s the total divorce rate has increased almost continuously. In the 1950s 13–15 per cent of marriages terminated in divorce, in the 1960s more than one fourth of marriages, and in the 1980s more than one third marriages were dissolved by the courts. In 1987, the legal procedure for divorce became more complicated in order to decrease the number of divorce. As it was expected, the total divorce rate indicated a sharp decline. However, the restrictive effects of the modification lasted for only one year. After a fall the number and rates of divorce have increased again.

Table 3
Selected indicators of divorces
1949–1994

Years, average of years	Number of divorces	Divorces per 1000 mid-year population	Total divorce rate (TDR)* per 100 contracted marriages	Mean duration at divorce** elapsed since marriage (in years)
1949	12,556	1.4	14.3	10.9
1950–1959	14,032	1.4	15.8	10.3
1960–1969	19,474	1.9	21.6	9.8
1970–1979	25,671	2.5	27.4	9.9
1980–1989	27,940	2.7	31.5	9.8
1990	24,888	2.4	30.9	10.1
1991	24,433	2.4	31.0	10.0
1992	21,607	2.1	28.0	10.0
1993	22,350	2.0	29.7	10.1
1994	23,417	2.3	31.8	10.3

Source: Volumes of Demographic Yearbook, Central Statistical Office, Hungary

* **Total divorce rate:** sum of marriage-duration-specific divorce rates based on divorces occured within 30 years elapsed since marriage per one (or 10, 100 etc.) of the contracted marriages.

** **Mean duration at divorce** is obtained by the weighted average of marriage durations at divorce, weights being the divorce rates at each marriage duration between 0 and 30 years.

The main role in the drastic increase of the level of divorces was played by the growth of the rates of marriages dissolved within a short period, namely within five years after the wedding. Divorce occurs most frequently in the second and third year of the marriage. Taking into consideration the necessary time for the appearance of conflicts between spouses and for decision on divorce, as well as the length of the time which is necessary for legal procedures, it can be said that in Hungary marriages get in a very short time into a deep crisis and they can be solved only by divorce.

Divorces of marriage cohorts

Comparing the marriage-duration-specific divorce rates calculated for periods with the corresponding indices of the real marriage cohorts it can be found that period effects play an important role in the pattern of divorces.

The Table 4 shows the decrease of the contracted marriages by divorce as a percentage by duration of marriage per year of marriage for Hungary. One of the features is that divorce rates are higher from cohorts to cohorts except for the youngest marriage cohorts whose behaviour may change a little. The other one is that the present and past behaviour within the same cohort seems independent from each other: divorce rates have increased in all marriage durations. Due to the radical increase of divorce in each marriage duration and in each marriage cohort the total cohort divorce rates have been close to those of period rates. The only difference is, that the total cohort divorce rates are slightly lower than the period ones.

It is obvious that not only legal regulations but also societal norms contribute to variations in the frequency of divorce. There are some important background factors increasing the incidence of divorce.

According to demographic and sociological studies, for example, early marriage means greater risk for divorce due to many circumstances. Early marriage can be used as a way out from family conflicts or disadvantaged backgrounds. Pre-marital conceptions are also frequent which are legitimized by marriage before birth. It was found in a survey carried out by the Sociological Research Institute and the Hungarian Central Statistical Office in 1980 for studying living conditions of broken families that more than one third of the respondents blamed a pre-marital pregnancy for the deterioration in their marital relationship.

Table 4
Cumulative proportions of marriages ending in divorce as a percentage by duration* of marriages for selected marriage cohorts

Year of marriage	After					
	5	10	15	20	25	30
	years					
1950	3.4	8.1	11.2	13.4	14.7	16.0
1955	5.6	10.8	14.4	17.0	18.3	20.0
1960	6.6	13.0	16.8	19.8	21.0	22.3
1965	8.1	14.8	19.3	22.6	24.1	25.4
1970	9.0	16.4	21.4	25.2	26.5	
1975	9.5	17.5	23.0	26.6		
1980	11.1	17.7	22.7			
1985	10.6	18.7				
1990	9.8					

* Duration in whole years on 31 December of the year of divorce.

Similarly, there is a higher risk of divorce where one or both parties have been married before. It often happens that the family life of those remarrying involves children coming from the earlier marriage of parties. The relationships within reconstructed families are complicated (e.g. stepfather, step-mother, old and new relatives etc.) and can make difficulties in adjustment.

Other explaining factors such as changing interrelations between the family and society, the role of women employment and the role of the two-income family model in family conflicts, the effects of the disappearance of rural society, financial and housing problems in families etc. have been also studied in Hungarian sociology.[6]

Some concluding remarks

The Hungarian population, compared with the trends of nuptiality in Western Europe, in general follows the European development. At the same time, in some phenomena it shows special features. In this respect, the social and political system determining the period following the Second World War, as well as the traditions in connections with marriage and family have played a significant role in Hungary.

The present paper has mentioned some factors contributing in the maintenance of the historical pattern of first marriage in Hungary. It should be also added that in Hungary, until recently marriage has been of high value in arrangements of family life. It has played an important role in the realization of personal independence and in the establishment of household. Additional factors are also the high level of premarital pregnancy and the parental encouragement to marry early, because parents usually support their children's marriages as soon as possible.

As far as the behaviour of the youngest population is concerned, the radical changes of the Hungarian society, including the appearance of new phenomena, such as the increasing level of unemployment, the spreading of individualistic values, an increasing propensity to live in cohabitation, etc should be mentioned. Although it would be difficult to predict the future, the present trends suggest a gradual withdrawal from the historical patterns of marriage. Especially, more children born outside of marriage confirm that important changes occurred in this respect.

Hungary was always known as a country where the proportion of illegitimate children was historically low. Indeed in the 1970s about 6 per cent of children were born out of wedlock. Since the second half of the 1970s the proportion of the illegitimate live births has been increasing and in 1994 their proportion has reached 20 per cent among births. This phenomenon is clearly in connection with the increasing popularity of cohabitation. However, at present we know little about this non-traditional form of family. According to the most recent statistical data for 1994, the proportion of population living in cohabitation has not been too significant yet, about 5 per cent of families consists of cohabiting partners with or without children.

Hungary has had a high divorce rate throughout the last decades. Under special circumstances divorce has become more rapidly an acceptable way of ending married life than in Western Europe. Thus the "golden age" was not a characteristic feature of marriage in Hungary after the Second World War as it was in Western Europe. Nowadays divorce does not mean social stigma and it may be less difficult for divorced persons to find a place in the society. A significant symptom of acceptance is that divorce rates have radically increased also in families with children. Although marriages without children are broken most frequently (only about half of them stand the time) families with two, three or more children show a radical increase in divorce. This may indicate that divorce is extending also to all strata of society.

Marriage and divorce behaviour have many "side-effects" affecting markedly the life course of individuals, the nature of family life, marital status and household composition of population. As far as the demographic consequences are concerned the past trends have played an important role in the composition of the present population by marital status. For example, the high proportion of widowed population has been in connection with their high propensity to marry in the course of their life. Similarly, it can be explained by the lower divorce rates of the earlier decades that the proportion of divorced population is not too high in the population. Present declining trends in marriage and the high level of divorce may have opposite consequences. Due to the close connection between marital status and household composition it can be expected that the proportion of people living alone in the population will be much higher in the future.

Notes

1. Jean-Claude Chesnais "Population trends in the European Community, 1960-1986". *European Journal of Population,* Vol. 3, 1987, 281–296.

    Hans-Joachim Hoffmann-Nowotny "The Future of the Family". In: Plenaries of the European Population Conference 1987: Issues and Prospects. Jyvaskyla. Finland. June 11-16, 1987, 113–200.

    Dirk J. van de Kaa "Europe's second demographic transition". *Population Bulletin,* Vol. 42, No. 1, 1987, 53.

2. John Hajnal "European marriage patterns in perspective" In: Population in History edited by D.V. Glass and D.E.C. Eversley. London Edward Arnold Ltd. 1965, 101–143.

3. Patrick Festy "On the new context of marriage in Western Europe". *Population and Development Review,* 1980, No. 2, 311–315.
4. Magdolna Csernák "Újabb tendenciák a házasodási viszonyok alakulásában". (Recent tendencies in nuptiality.) *Demográfia,* Vol. XXXVII. 3–4, 1994. 298–314.
5. Magdolna Csernák "Az első házasságkötések alakulása Magyarországon a II. világháború után. Születési kohorszok házassági táblái". (First marriages in Hungary after the World War II. Nuptiality tables of birth cohorts. Publications of the Demographic Research Institute.) NKI Közleményei. 1983/1. No. 54. 295.
6. László Cseh-Szombathy "A házastársi konfliktusok szociológiája". (Sociology of the conflicts within marriage) Gondolat Kiadó, Budapest, 1985. 158.

Marietta Pongrácz, Magdolna Csernák "Divorce in Hungary". In: Later phases of the family cycle. (Ed. by E. Grebenik, C. Höhn, R. Mackensen) Clarendon Press, Oxford, 1989. 37–54.

Ferenc Kamarás

Birth Rates and Fertility in Hungary

A Period of Demographic Transition from the Late Nineteenth Century to the Second World War

At the present time in Europe demographic trends are marked by a decline in mortality and fertility rates. However, the European countries significantly differ in timing and duration of these periods, and in the way they took place. The period of this change has a stage when mortality improves significantly, but fertility is still high, which gives rise to a fast and remarkable population growth. In Hungary this period was rather short, and in the first decades of our century also fertility fell significantly (Diagram 1).

The rate of forty-five live births per thousand in the 1880s could be considered high under the demographic conditions of the age and fell below forty per thousand only slowly around the turn of the century. The slow decrease continued in the first decade of the century, until the First World War left its deep and long-lasting impact on the development of the population. During the war, fertility fell drastically involving temporary reduction in the number of the population. Between 1915 and 1919 the number of the newborn babies was 500,000 less than in the previous five years, which meant that during the war years the fertility of nearly two full years was missing, and the reduction in the number of births reduced the number of the population by two full generations. The full strength of the generations born between 1915 and 1919 was 40 per cent less than that of those born five years earlier and 36 per cent less than that of those born between 1920 and 1924. This wave-trough of births left its impact on the age-pyramid for the decades to come. Even if the fertility level had remained unchanged, these female generations small in number could only have produced fewer children. But even fertility fell drastically in the mid-war period. Rapid decrease started in the years of the world economic crisis, i.e., in the early 1930s, but the smaller number of births at the end of the decade and in the early 1940 could be attributed also to the smaller number of women in their productive age born during the First World War.

Fertility conditions and the frequency of childbirth can be traced in present-day Hungarian territory only from the turn of the century. The total fertility rate indicates the possible number of children of a woman, should the frequency of birth in a given year of period apply to her, i.e., the conditions of reproduction in a given population. In spite of the decrease in the number of births in the late nineteenth century, the average family at that time had more than five children (5.3), which ensured the simple reproduction of the population even under the circumstances of high infant mortality. By 1910, there were only 4.7 children in a family, then between 1910 and 1920 the pace of decrease accelerated and was doubled as compared to the previous decade. The years 1920 and 1921 seemed to wish to compensate for the losses during the war years and produced higher fertility. The total fertility rate still lagged behind the one ten years earlier by twenty-three per cent, which meant an average of 3.7 children for a mother. A decrease of a similar rate continued also in the 1920s, and in 1930 the total fertility rate was hardly more than half of the one around the turn of the century (2.8). It is interesting to note that in the twenty years between 1910 and 1930 the rate of fertility decrease was nearly constant and was the highest in Hungary, provided one leaves the extraordinary period of the First World War out of consideration. Between 1930 and 1940 the level of fertility went on decreasing, but at a slower pace, comparable to the situation in the first decade of the century. All in all, by 1940 the total fertility rate of the Hungarian population had fallen to hardly more than half of its rate around the turn of the century (2.5), i.e., it decreased by 13–14 per cent every decade (see Table 1).

This significant drop in childbirth did, however, not characterize all age-groups of women equally. It was the older age-groups of women in their productive age that showed lower fertility rates, which meant that women had fewer children of higher parity. Women under twenty played only a small part in the decrease, and the fertility of women under seventeen even increased by fifteen per cent in the first decade of the century. Women between twenty and twenty-four showed a decline of only five or six per cent, and it was only in the case of the generations above twenty-five that the decrease was above ten per cent. In those years the tradition of early marriage and first child soon after was still prevalent. In the following decades decrease could be felt in all age-groups, though to a different degree. The dividing line was remarkable between women above and under thirty. Between 1910 and 1920, the fertility of women in the age group 20–29 fell by 18–19 per cent, while that of those between 30 and 39 fell by 27 per cent. The level of fertility of women under twenty did not change significantly between 1920 and 1940, and even started to rise moderately in the following decades. The fertility of women above thirty and even more of those above thirty-five was falling at a constant pace in this period, too, and showed a decrease of 20–25 per cent in each decade.

To sum up, during the first forty years of this century the fertility of women under twenty fell by around 30 per cent, that of women between 20 and 29 fell nearly to half of the previous figure, and that of women above 30 fell to about one third of the figure around the turn of the century.

Fluctuations of Fertility from the Second World War to the End of the 1970s

The Second World War did not produce as deep an effect on fertility as the First World War, but mortality rose significantly because of the war itself. After the war, there was no spectacular rise in fertility to make up for the losses, but the number of births was steadily rising until the early 1950s, when it began to fall, giving rise to political intervention. The prohibition of abortion resulted in a radical rise in the number of births. These strict measures raised fertility by 20 per cent in the following two years, and in the case of the older generations of women even by 25 per cent. This trend of forceful administrative measures lasted, however, only a few years and was lifted in the second half of the 1950s. Interestingly, one of the most liberal systems of birth control followed, making abortion possible for nearly everyone who asked for it. Since modern contraceptives were not yet available in Hungary, the number of abortions started to increase rapidly. So the subsequent fall in the number of births could partly be attributed to unlimited abortions. However, the actual cause was far beyond this, since abortion was an effect in itself, an unwanted result of a diminishing inclination to give birth. The real causes were more complex.

The social and economic changes beginning in the second half of the 1950s and accelerating in the 1960s influenced both individual and family life, including the number of children in a family. The forced process of industrialization made large masses of former housewives wage-earners, and even commuters. The elimination of individual farming created large-scale territorial and social mobility from village to town, from agriculture to industry. Female employment involved higher qualification for women, so women had to devote more time to education. At the same time, there was a change in the position of the family within Hungarian society, and the role of children within the families changed, too.

The drastic fall of fertility reached its nadir in the early sixties, when Hungarian fertility rates were the lowest in the world. The number of those born in 1962 was 100,000 less than that of the generation born in 1954. The total fertility rate fell from 3.0 to 1.8 in the same period, which meant a forty-per cent fall in fertility concerning all age-groups. The absolute number of live births (130,000) was hardly higher than that in 1918 (128,000), which was a historical minimum during the First World War (see Table 2).

This perennial low level of fertility foreshadowed an unfavourable turn in Hungary's demographical situation leading to population aging, a decrease in the total population, along with socio-economic problems arising from the unfavourable age distribution of the Hungarian society. Given the system of two wage-earners in a family and nearly full female employment, most children were born of working mothers. It became therefore obvious that the women's inclination to have more children could be increased only by incentive measures and by reducing the tension between motherhood and professional life. Upon such considerations the child-care allowance called GYES was introduced in 1967 making it possible for working mothers to remain at home with their children on a regular benefit until the children reached the age of three. When initiated, this benefit amounted to nearly a quarter of the average income of a woman. The mother remained on the pay roll and regained her job when the three years were over. This system also served to relieve the most expensive institutions of child-care, the overcrowded infant nurseries.

The child-care allowance became very popular, and in the beginning more than ninety per cent of working mothers took advantage of it. Its introduction led to a slight increase of the number of births, and the total fertility rate also increased a bit as compared with the nadir in the early 1960s. Especially the younger generations undertook to have more children, and their fertility kept rising from the second half of the 1960s to the late 1970s.

To keep the strength of the population on a given level for a longer period, which is called simple reproduction, and to create healthy age distribution in society, the total fertility rate should have been around 2.1–2.2.

However, because the introduction of the child-care allowance failed to help reach this level, in 1973 new measures were taken to support demographic growth, which aimed to attain levels of simple reproduction. As the government wished to encourage people to have more children, they applied financial, pursuasive, and legal means to attain their goal. Besides these incentive measures, abortion was also restricted, and modern contraceptives were made available.

In the two years after 1973, the level of fertility rose by a significant twenty-two per cent. At the time of the introduction of the above-mentioned measures, the numerous generations born around twenty years earlier were entering their productive age, and this contributed to the significant rise of the number of births. Especially second and third children were born in greater proportions. Since both incentives and penalties comprised the legislation, it is uncertain which contributed more to the increase. Women were allowed to have an abortion only after two live births, but families of two wage-earners and two children had become general anyway, and most women wanted to have two children. The number of abortions fell significantly, but the number of rejected applications did not rise very much as compared to the previous period. Although we can establish that most children were desired ones in the mid-1970s, it was just that their parents might have preferred having them later. The improved allowances provided by the child-care benefit made it easier to finance a home, which contributed to the birth of a first, second or even third child in a shorter period.

However, the rise of fertility rates did not prove lasting, and the level of simple reproduction was surpassed only for a few years. Fertility started to fall again from 1976, and by 1980 it was lower than in 1973, i.e., before the introduction of the encouraging measures. So the new wave of fertility culminated in 1975, twenty years after the prohibition of abortion in the mid-1950s (see Diagram 2).

Characteristics of Fertility in the Past Fifteen Years

The protracted, basically downward tendency of fertility is a well-known fact in developed societies. In Hungary, this process took place in the same way as in Western Europe. However, fertility below the level necessary for maintaining status quo in population over a long period of time is a relatively new phenomenon in Europe, but it is getting to be the normative demographic characteristic of developed societies. It involves the aging of the population along with a lasting, significant fall in the gross population over a long run. Apparently it is no temporary phenomenon, because the social and economic factors underlying it seem to be permanent. The decrease of fertility is therefore a demographic response to the changes in society inflicting primarily the family as its fundamental unit. Its concomitant circumstances are the decrease in the number of marriages, the increase in the number of common law partnerships, childbirths out of wedlock, and the decreasing stability of partnerships of both kinds. These phenomena became widespread in Western Europe in the 1970s, and serve as examples to be followed also for developed societies overseas.

Characteristics of Period Fertility

Western European demographic patterns appeared in Hungary as early as the beginning of the 1980s, and became more and more widespread in the 1990s leading to fertility patterns quite different from previous ones. As regards the level of fertility, Hungarian demography anticipated Western European societies, for numbers first decreased in the years after the Second World War and fell below the level necessary for simple reproduction. This process is ongoing. There were, however, certain aspects of fertility that differed from the Western European patterns. In Hungary marriage at young age involved childbirth at young age, intended childlessness was rare, every woman wanted at least one child, families with two children became the norm, childbirth above thirty or thirty-five was rare, and few children were born out of wedlock.

This pattern tended to break up in the early 1980s, and the change has recently assumed epidemic proportions. What changes are they? The protracted downward tendency of fertility took place parallel with mothers' getting younger. In other words, it was the fertility rate of mothers above thirty and the births of children of higher parity that were decreasing, while those of younger mothers even started to increase. However, from the early 1980s, the downward tendency went hand in hand with the aging of the mothers for the first time in this region. The fertility of those under twenty fell to less than half of the 1980 level, while that of those between 20 and 24 decreased only by thirty per cent. At the same time, the fertility of mothers above twenty-five was still rising until the mid-nineties (see Diagram 3). As a result, the rate of children born to mothers between 25 and

40 was still twelve per cent above the fertility of the same age-group in 1980. So we are witnesses of a significant structural change in fertility, as is shown by the comparison of age-specific fertility rates in 1980 and 1994.

The graph shows a fifteen per cent reduction in the level of general fertility since 1980, for the moderate rise of fertility of women above twenty-five could not compensate for the smaller fertility of younger women. At the same time, the modal value of the fertility curve rose from twenty-one to twenty-five, which means that women generally have their children four years later than before (see Diagram 4). These structural changes have altered not only the modal value of age-specific fertility rates, but also the average age of mothers at the birth of the first child. The modal value has been steadily rising since the early 1980s, and the change is most conspicuous in the case of married women, who are generally 1.2 years older now at the birth of their first children than fifteen years ago, which comes from the fact that women tend to get married later. Not only married, but also unmarried mothers give birth later nowadays, though this rising tendency has slowed down a bit since the second half of the 1980s. This phenomenon goes back to the shift to births out of wedlock, which is primarily a factor among younger unmarried mothers (see Diagram 5).

The shift towards older age-groups involved significant changes also in the age distribution of live births by birth order. The average value expresses the birth order of a child born to a given mother. Hungarian fertility conditions have long been characterized by the dominance of first and second children. The general decrease of fertility involves the higher rate of first and second children among all live births. They constituted 85 per cent in 1980, and the mean of birth orders was 1.82. Falling fertility rates were coupled with a rise in the mean of birth orders for the first time in the mid-1980s. The rate of first and second children decreased owing to the smaller fertility of young women, which automatically emphasized children of higher parity, raising by this the mean of birth orders. Its value in 1994 was already 1.95.

The changes of fertility by parity is quite similar. This value expresses the fertility of mothers with one child, and those with two or three children. Because calculation of such values has become possible only recently, it is worthwhile introducing the trends for a longer period, i.e., from the early 1970s. Diagram 6 shows the fertility of women without children and of those with one child only. In the early seventies mothers were fairly young at the birth of their first child, and many women gave birth at a young age. One of the causes of this was that 20–25 per cent of the brides were already pregnant when the wedding took place. Childbirth out of wedlock was not yet general, so the children had to be legalized by marriage. The measures promoting childbirth in the mid-1970s also contributed to the birth of first children, but even more encouraged the birth of second ones. The fertility of mothers with one child temporarily exceeded that of childless women. Lasting changes came with the mid-1980s when the steadily falling tendency of fertility among childless women accelerated, and second children were born to much higher proportions than first children. Fertility rates by age and parity make it obvious that childless women of twenty-five and below undertake to have their first children to a much smaller extent than in the early 1980s (see Diagram 7). The fertility of childless women above twenty-five has, however, not changed at all, and the same can be said of mothers with one child. The cause of the smaller rate of second children in these days as compared to 1980 lies in the lower fertility of mothers with one child under twenty-five, since the rate of second children in higher age-groups has not changed at all. These trends go back also to the significant changes of attitude as regards marriage mentioned before.

The changes of fertility of mothers with two and three children are also remarkable (see Diagram 8). The fertility of mothers with three children steadily exceeds that of the other group. The only exception was the middle of the 1970s when this was not the case. The measures introduced at that time had an effect mostly on the fertility of mothers with one child and two childen, and contributed to the birth of second and third children to a greater degree than before. The middle of the 1980s means a dividing line here, too, for the downward tendency stopped and started to rise. As a result, the fertility of mothers with two children is the same now as in the early 1980s, and that of mothers with three children is 10 to 15 per cent higher. The fertility of mothers with four and five children is also surprising, since it has been steadily rising since the mid-1980s. What is the explanation for the rising tendency of fertility of higher parity since the middle of the 1980s? It is probably right to say that GYED, another kind of child-care benefit called child-care fee and introduced in 1985 contributed to the positive changes. (GYED entails that mothers bringing up their children at home get 65–75 per cent of their former salary for two years, a sum greater than the earlier fixed sum taking no account of the mothers' income. The primary aim of the introduction of GYED was to promote fertility in all layers of society, to diminish the differences by social status, and to recognize the social value of childbirth also in older age-groups and in the case of mothers with a higher income.)

The change of fertility rates by parity throws new light upon fertility conditions in Hungary today. Since the fertility of childless women falls back on the long run and to a considerable degree, the rate of women with one child is reduced, and consequently the rate of potential mothers with two children. The actual rate of mothers with two children has decreased only slightly since the mid-1980s and can be said to be stagnating. As a result, the rate of mothers with two children and of those potentially having several children also decreases. Where a second and third child is born, there is now a greater possibility of another brother or sister than in the beginning or in the middle of the 1980s. The same applies even more to mothers with four and five children. So the fertility of mothers with several children suggests good perspectives for the future, while the general level of fertility is still decreasing. The reason for this lies in the rate of mothers with several children. However fertile they may be, their rate among mothers is diminishing, and fertility conditions in Hungary will be determined by the rate of first and second children, if any. For the time being, these children are either not brought to life or only later, which is very different from the conditions in the early 1970s or in the mid-1970s characterized above. So we are witnesses of a strategic change as regards marriage and childbirth, the short-term effects of which are becoming manifest nowadays.

Another new element is the growing rate of childbirth out of wedlock, also indicating a change in behavior and norms in the past decade. The rate of children born out of wedlock has reached unprecedented proportions in this century. In 1994 the rate of children born out of wedlock was 19.6 per cent, which is two and a half times higher than in 1980, and three times higher than in the mid-1970s. There are significant territorial differences Budapest having every fourth child born out of wedlock. Seventy-seven per cent of the mothers concerned were single and twenty per cent divorced. The rate of single mothers has been rising drastically in the past few years, and from the early 1990s their fertility surpasses that of divorced mothers (see Diagram 9) in spite of the fact that their rate within the population is also growing. The decline in incentives for marriage made the rate of single women rise, and the lower rate of remarriages contributed to the higher rate of divorcees and widows, whose rate even among women in their productive age is also increased by high male mortality.

The distribution of children born out of wedlock by the age-groups of the mothers is also very interesting (see Diagram 10). The younger the mothers, the more children are born out of wedlock. Nearly one third of all children born out of wedlock belong to teenage mothers. Cohabitation might be general among these mothers. However, our data concerning their way of life are still sporadic. A representative survey conducted in 1993 shows that adolescent cohabitation with the first partner is three or four times more frequent now than about twenty years ago, in the young age of the mothers. Nearly forty per cent of these young mothers (age-group 18–19) had no steady partners at the birth of their first babies. So childbirth out of wedlock seems to have lost its role of enforcing marriage with the father. Women undertake to have children even at the expense of having to bring them up without a father. Forty-three per cent of children born to mothers under twenty are born out of wedlock, which is 2.3 times more than in 1980. Owing to the high rate and young age of this group, it can be considered the most serious social problem, especially if marriage does not follow. As mothers get older, the rate of children born out of wedlock gets lower. It is the lowest in the case of middle-aged mothers, since the majority of women in this age-group are married. However, once a middle-aged mother is not married, the rate of children born out of wedlock is growing more rapidly than in the younger age-groups. The increase in the age-group 20–24 was threefold, and among those between 25 and 29 it was 2.5 times higher than in 1980. In other words, in age-group 20–24 the rate of children born out of wedlock grew from 5 per cent in 1980 to 15 per cent, while in the age-group 25–29 it grew from 4 per cent to 11 per cent. If these children were born in cohabitation or brought up without a father is not precisely known.

Changes of Completed Fertility

Completed fertility is considered to be the "purest" index of fertility in the sense that timing during the productive age does not fluctuate fertility any longer, as it does in the case of period fertility. The final number of children does not reveal the actual fertility conditions of a given period, but expresses the total fertility of a generation in its productive age. The problem is that researchers have to wait at least twenty-five years to be able to interpret the data properly, even though the productive period is getting shorter. The continuous time sequences of completed fertility can also be introduced. So it is justified to offer a long-range survey starting with 1970 here, too. Since mothers above forty represent only hardly more than one per cent of the total number of births, the number of children born to mothers of forty can be considered final.

Diagram 11 shows the completed fertility of the last age-group of women in their productive years, i. e., those between 45 and 49. The average number of children in the age-group was significantly and steadily falling quite up to the second half of the 1980s, but it stagnated afterwards. It is interesting to note that the strict prohibition of abortion in the 1950s that resulted in a radical increase of the number of births on the short run did not influence the completed fertility of the mothers, since the decrease was unbroken. Stabilization became manifest mainly in the case of the generations born between 1940 and 1945, i.e., in the case of those entering their productive age in the late 1950s or early 1960s. In other words, the recent measures trying to encourage mothers to have more children effected the women whose productive years have just come to an end only to a degree that the rapid downward tendency could be stopped. At the same time, the average number of children (1.85 to 1.90) of mothers between 45 and 49 was the lowest in the period in question and was below the level necessary for simple reproduction by about fifteen per cent.

The situation is different with the fertility of the age-group 40–44 that can be considered quasi finished productivity (see Diagram 12). The downward tendency was significant here, too, quite up to the early 1980s, when the fertility level got stabilized first, then a rising tendency began. The fertility of the age-group born after the Second World War is about ten per cent higher at the present than that of women of the same age but born during the war. When this generation was young, the child-care benefit system was introduced; when they were in the middle years of their productive age, they were effected by the measures of 1973, then also by the introduction of the child-care fee (GYED) in their later years. The generations that were forty to forty-four in the first half of the 1990s had finally more children than the similar generations earlier or the age-group 45–49 ahead of them. Their average number of children (1.95) is still ten per cent below the level necessary for simple reproduction. The stabilization or slight rise of completed fertility in some age-groups is due to the fact that the rate of childless women has been falling, the rate of mothers with two children has been rising significantly, and that of mothers with three or more children has not been changing since the second half of the 1980s.

Fertility as Reflected in Surveys of Family Planning

Fertility and family surveys have a relatively long history in Hungary. The following summary is based on the results of three longitudinal marriage surveys and the cross-sectional survey of 1993. These surveys were nationally representative and referred to a given cohort or a given sample of women in their productive age.

We followed the fate of couples married in 1974 for sixteen years between 1974 and 1990. The size of these families in 1990 can be considered to reflect finished fertility, so the original plans of the couples can be contrasted with the actual number of their children. The lessons to be drawn are the following:

- Intended childlessness is not typical in Hungary as yet. Almost all couples wanted at least one child when they got married, and the rate of childless women remained on the level of infertility due to biological reasons (4 per cent).

- Families with two children were typical both in the plans of the couples (72 per cent) and in reality (57 per cent). Although the actual rate of families with two children remained below the expectations, it has never been so high. Many who originally wanted two children had finally only one, so the planned rate of single children (6 per cent) turned out to be much higher (19 per cent).

- Families with three or more children represented 20 per cent, which is only slightly less than the planned rate of 21 per cent. This is very positive, and can be attributed to the measures encouraging families to have more children to a great extent. The rate of families with three children is somewhat below the planned figure, it is still unusually high as compared to the figures of finished fertility in the past few decades. The rate of families with four or more children was, however, more than intended. Still, their rate has never been as low as that.

It is interesting to compare the actual number of children in a family with the number held ideal by the parents. The difference is the greatest in the case of families with one child and with three children. Many more families have only one child and far less have three children than the parents would find ideal for themselves. Six or seven per cent of the families planned to have one child and held it ideal, while the actual rate of families with one child was three times as much. In some cases it was due to biological reasons like the so-called secondary infertility, health problems, or the disintegration of the family, and divorce. At the same time, 28 per cent of the families found three children ideal, 19 per cent planned to have three children, but only 16 per cent had actually three children when the productive years were over. The causes of the difference may be financial problems, difficulties with harmonizing work and family, and the lack of a proper home for such a big family.

As a matter of fact, the interviewed women expressed their opinion about the ideal number of children sixteen years after their weddings, at the end of their productive years. Thinking about their lives in retrospect they may have idealized the number of children. Circumstances had made it impossible for them to realize their plans, let alone the ideal size of their family. The situation was quite the opposite in the case of families with four or more children. Four per cent of the families belonged to this category, while it was only two per cent that actually planned it, and two per cent held it ideal to have so many children. In some cases the cause might have been the lack of family planning ("I got pregnant and we kept it"). All in all, the actual number of children born to couples getting married in 1974 was eight per cent less than what had been planned before and ten per cent less than the number of children they held ideal afterwards.

Another longitudinal survey followed couples getting married in 1983, who have not reached the time of completed fertility. Their marriages in the first six years were monitored. Their plans fell short of those of the couples of 1974, but the average was still more than two children. This average decreased only slightly in the first six years of their marriage, owing to the disintegration of some marriages. Sixty per cent insisted on the plans, twenty-four per cent wanted fewer children and sixteen per cent more than they had planned before. The majority of those who changed their minds wanted only one child instead of two, and more than half of those who had wanted three children were satisfied with two as years went by. However, it is interesting to note that the rate of couples planning three or four children increased. The majority of these couples came from the group who had planned two children but later wanted more. As a result of these changes, the rate of those planning one child rose from 9 per cent to 18 per cent, the rate of those wanting two children fell from 75 per cent to 63 per cent, and that of couples wanting three or four children rose from 15 per cent to 17 per cent. So the average number of children planned by the couples was 2.00 six years after the marriage in contrast with the average of 2.07 at the time of the marriage.

The third longitudinal survey included young couples getting married in 1991. In their case data were collected only once, so only their plans in the first year of their marriage can be analyzed. The average number of planned children was 2.12, which is near the level necessary for simple reproduction. It was especially interesting that 23 per cent of them wanted three or more children, which exceeded the similar rate in the former surveys. Sixty-one per cent wanted two children and there was no sign of intended childlessness (see Table 3).

The cross-sectional survey represented women of 18 to 41 on a national basis. This was the first survey that covered all women irrespective of their marital status. The number of the children these women wanted was also relatively high, an average of 2.10. Young women under twenty-five wished to have the most children (2.15), but the average was decreasing with the older age-groups. Women of forty or forty-one, approaching the end of their productive years, wished to have less than two children (an average of 1.99). This figure can be considered final, since the overwhelming majority of the desired children had been born by that time (1.90). Both the longitudinal and the cross-sectional surveys reveal that as years go by, and with the advancing years of age women tend to wish to have fewer children, and even these reduced plans are not always fulfilled. Families with one child will finally be much more numerous than planned. Although only ten per cent of the young women planned to have one child, their rate was never below twenty per cent even by a finished fertility much higher than today. The other reason of the decrease is that while 65–70 per cent plan to have two children, the actual number of women with two children is not above 55 per cent when their productive years are over. The desired level is not reached even today, though the rate of women with two children has never been so high before.

The results of surveys covering family planning are often received sceptically even by the experts of the subject. The desired final number of children in a family consists of the number of the already living children and of the ones planned for the future. So it is questionable if this index can reflect the prospective fertility conditions. The higher the rate of the still planned children as compared to that of the already living ones is, the more uncertain the index is, since couples may change their minds as to the final number of their children. Uncertainty is more pronounced with the younger generations and with single women than with older generations towards the end of their productive years.

As regards the fertility trends of the 1990s, the survey results, especially the ideas of the young about their future families, are noteworthy. The fact that young people tend to live in partnership without getting married and tend to have fewer children than before in young age does not mean that they want to have fewer children in their future families in the end, which is a positive phenomenon. It is, however, also a fact that the difference between the planned and the actual number of children tends to be ever greater. Given the fertility conditions for 1994, a woman could have an average of 1.64 children in her lifetime, while the desired number of children is more than two, and the actual number at finished fertility is usually also around that figure. This means that timing is different now, and young couples put off having children until their living standards are higher. While in the 1970s the significant increase in period fertility measures was due to childbirth in young age, it is the opposite trend that is decreasing them now.

Having children later in life is not negative in itself, if intended timing is involved. Childless years help partners prepare for responsibility and determine the roles within the family. Some experts say that these years even contribute to the stability of the marriage. The demographic consequences are, however, not so unambiguous. Experience mainly from Western Europe shows that having children at a later date involves having fewer children at the same time. The rate of families with one child or without children may rise as a consequence. The intentions of the would-be parents will not necessarily come true. Changes in family policy nowadays make it uncertain if the young people of today would have two or three children in proportions similar to their parents. The signs are not at all encouraging. The year 1995 is another nadir in Hungarian fertility. The assessed figure of 112,000 live births is the lowest in the past hundred and twenty years. The total fertility rate has fallen to 1.57 and the gross reproduction rate of around 0.7 forecasts the strength of the generations of children growing up under such fertility conditions thirty per cent smaller than that of their parents. Taking period fertility by parity into consideration, one can draw conclusions as to the distribution of completed fertility by the number of children. Results covering the early 1990s reveal that should the given fertility relations by age and parity be lasting, the rate of childless families would double, that of families with one child would increase, and that of families with two children decrease in the future. This would have a negative effect on our demographic conditions, since the level of completed fertility would fall and lag behind the figures necessary for the simple reproduction of the population. The spread of Western European fertility patterns alongside Eastern European mortality would lead to a significant decrease of the population on the long run.

The Situation Abroad

Fertility trends in Hungary and in the rest of Europe in the past decades do not forecast a lasting rise in fertility reaching the level necessary for simple reproduction. Sweden is the only exception where fertility was among the highest in the early 1990s and temporarily reached and even surpassed the above mentioned level. This was caused by a deliberate state policy designed to allow women to work as well as to raise children through providing allowances similar to the Hungarian child-care fee. In the 1990s it was only in Denmark and Finland that fertility rose slightly. In the other European countries the former downward tendency continued. Decline was considerable or even dramatic in the former socialist countries called also "transitional societies". The situation in former East-Germany is the most critical where fertility fell to half (!) of the previous level between 1990 and 1994, i. e., following the unification of the country. Decline was around thirty per cent in the newly independent Baltic states, fertility fell by a quarter of the former figure in the Czech Republic and Romania, and by one fifth in Bulgaria, the Slovak Republic and in Ukraine. The fall of the level of fertility after the change of political regimes was the smallest in Poland and in Hungary and, strangely enough, Hungary had the second highest fertility rates in 1994 among the former socialist countries, while prior to the political change Eastern Germany and Hungary had been the last in the row. The slower rate of decrease might be attributed to the fact that in the years between 1990 and 1994 the former institutions and measures promoting families, mothers, and children remained unchanged and were even extended.

Out of the thirty-five countries of the Council of Europe it was only Cyprus, Iceland, and Turkey where fertility reached the level necessary for simple reproduction in 1994, whereas twenty-five countries reached this level in 1970. Even then Hungary was not one of them. The authorities of the different countries reacted to the lasting low level of productivity differently. According to government reports from 1990, none of the thirty countries of Europe considered their fertility rates high. Nineteen countries were dissatisfied with their fertility level, with Spain and Austria among them, where the total fertility rate was below 1.5 at that time. However, the number of countries considering their fertility rates low increased. Eleven countries found their fertility level low and, with the only exception of Germany, all of them considered state intervention necessary. These countries were Bulgaria, Hungary, Romania, Greece, Italy, France, Lichtenstein, Luxemburg, Monaco, and Switzerland (see World Population Monitoring 1991). Greece and Italy had a total fertility rate below 1.5. Since then, the situation has changed significantly for in 1994 already twelve countries had such a low total fertility rate and the population of ten countries was diminishing.

Hungary is ahead of these countries in demographic sense, for the low level of fertility below the one necessary for simple reproduction has been going on for the longest period, and the number of the population has been decreasing for fifteen years running, and to an ever growing extent. To ensure simple reproduction is of fundamental importance from demographic point of view, for both aging and the further decrease of the population cannot be stopped without reaching this level. Let us leave international migration and its possible demographic consequences out of consideration for the time being. Not because we do not consider the subject topical, but only because there is no example for migration solving demographic problems anywhere in the world. They generally depend on the economic and political situation in a country and change according to relevant interests.

It cannot be established if the social, economic, environmental, and mental factors working against families' having more children and leading to a long drawn-out decrease of fertility have diminished or not. The future cannot be foretold. All population forecasts speak of a lasting decrease going beyond the turn of the millennium. Especially the number of the children and the younger generation will decrease, while that of the elderly people will grow significantly. Besides the aging of the population, the upset balance of the economically active and inactive part of the population will create further social and economic problems resulting from the growth of the number of dependents, the survival of the old system of provision for pensions, and the health and social welfare problems of the old. These problems are manifest already by the present distribution of the population and nowhere in the world have practical solutions been found to solve them. There are no examples or beaten tracks to be followed. While the developing countries and the world as a whole are justly worried about overpopulation with all its consequences, developed societies have to face a different challenge resulting from their demographic conditions. From demographic points of view Hungary belongs to the latter category with the difference that the same problems must be solved against a much less favourable economic background.

Diagram 1
Live Birth and Mortality Rates, 1900–1995

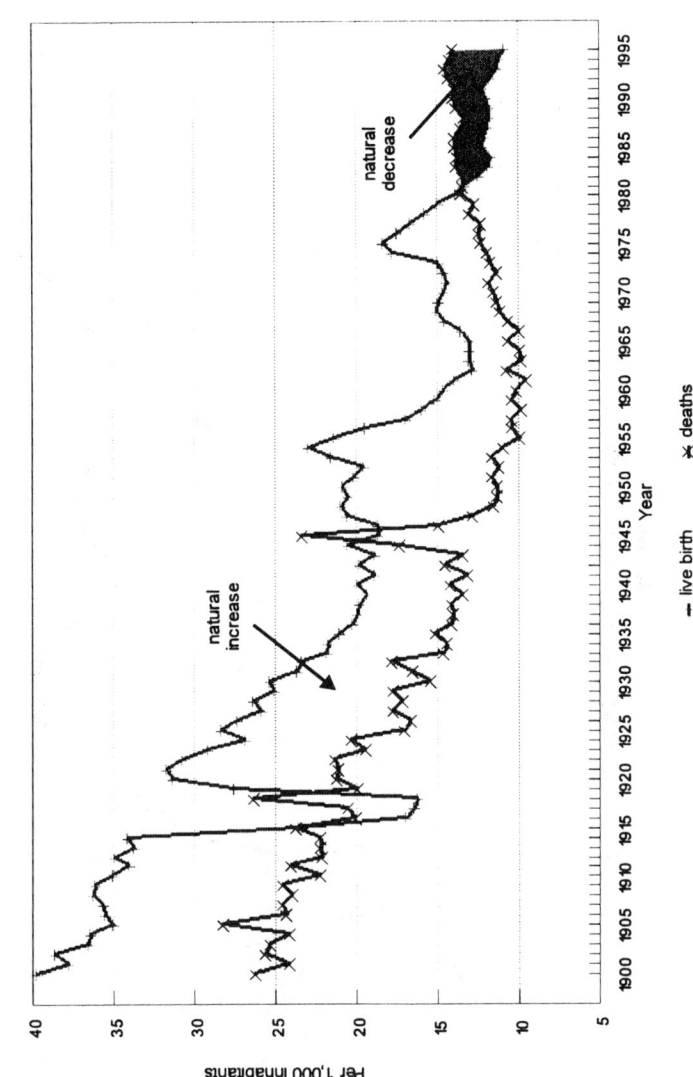

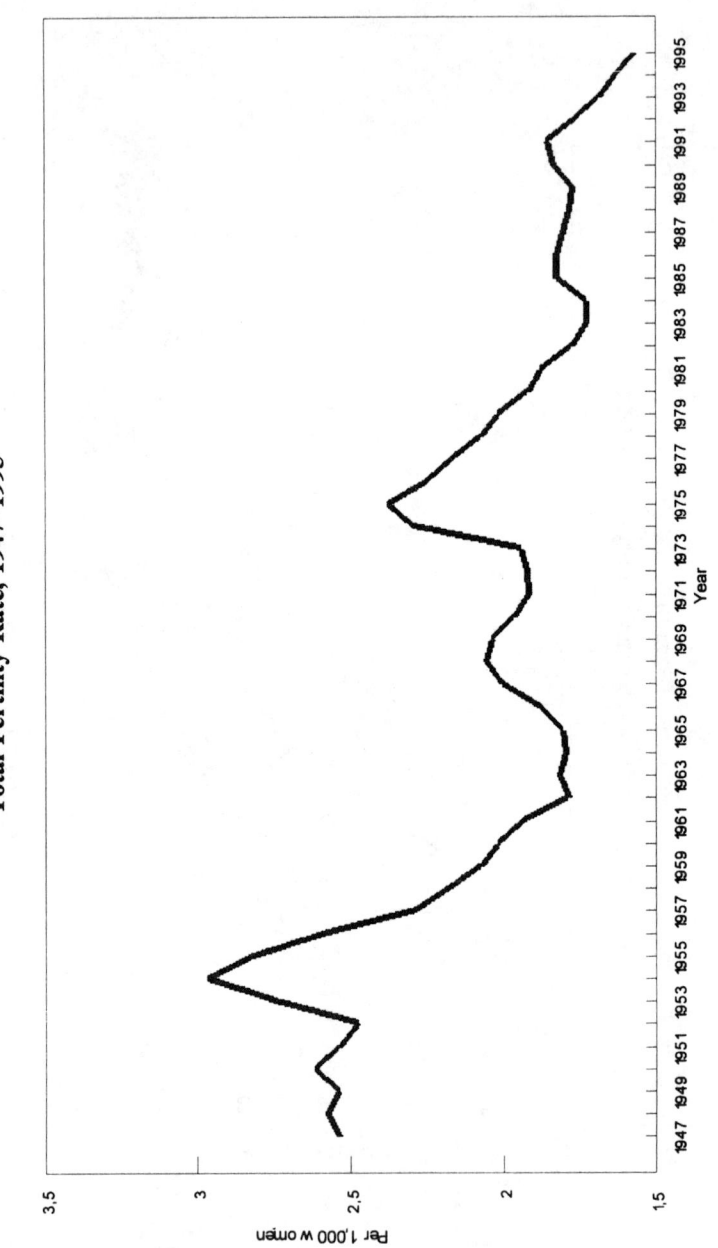

Diagram 2
Total Fertility Rate, 1947–1995

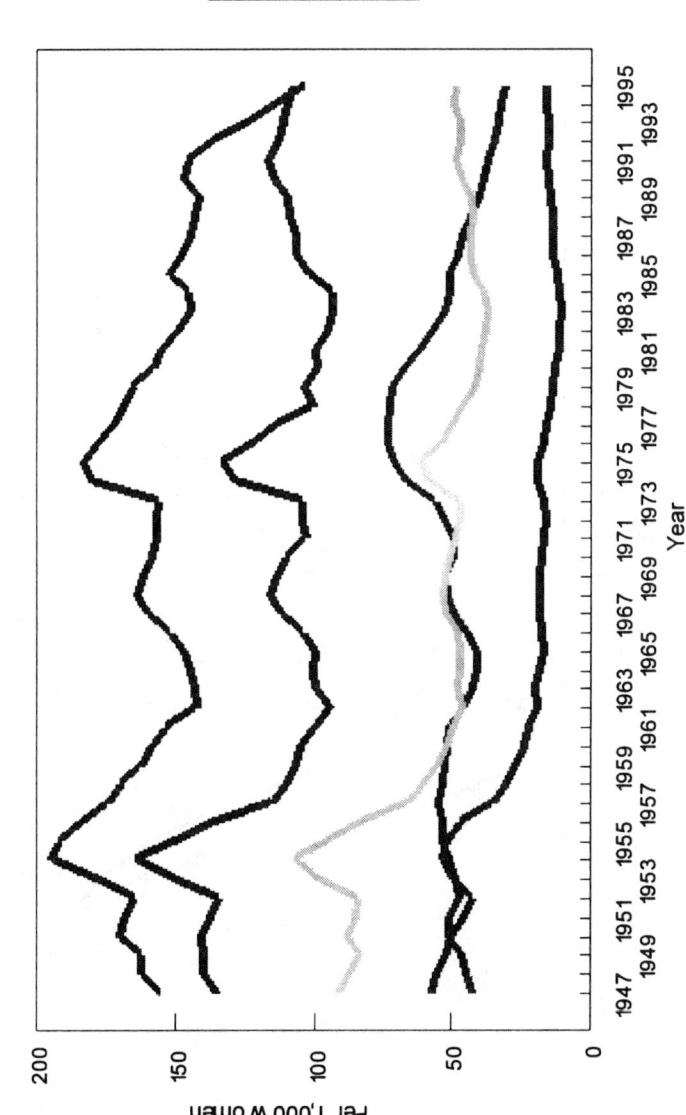

Diagram 3
Age-specific Live Birth Rate 1947–1995

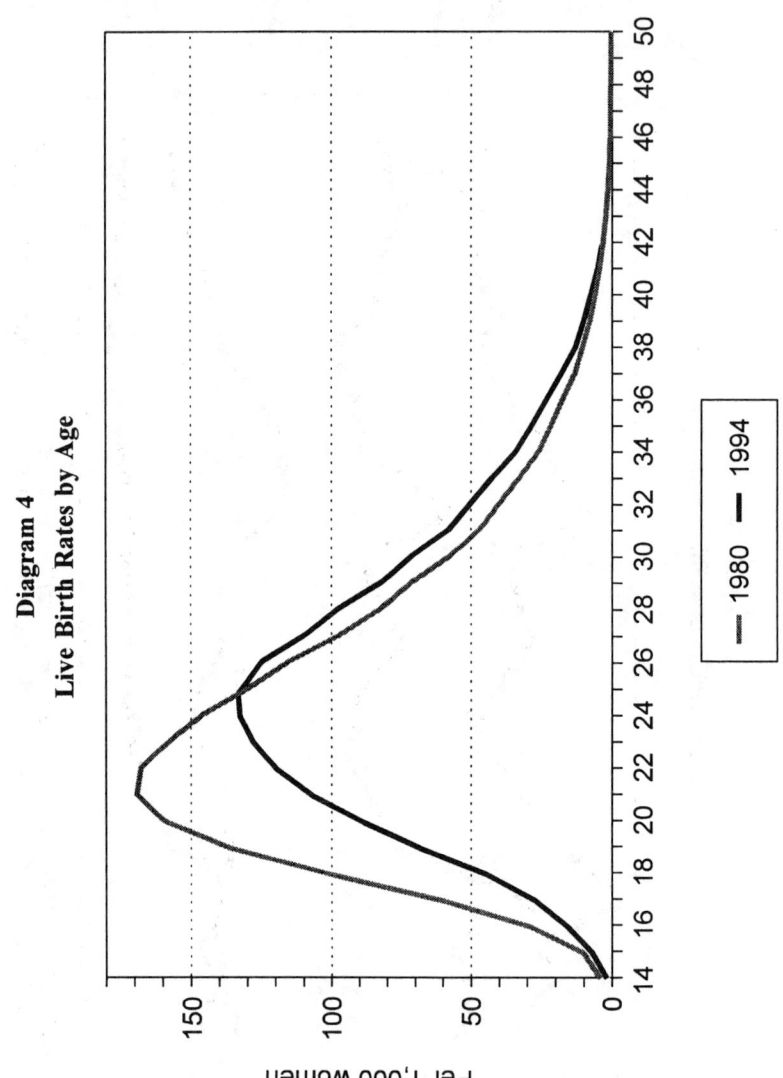

Diagram 4
Live Birth Rates by Age

Birth Rates and Fertility in Hungary

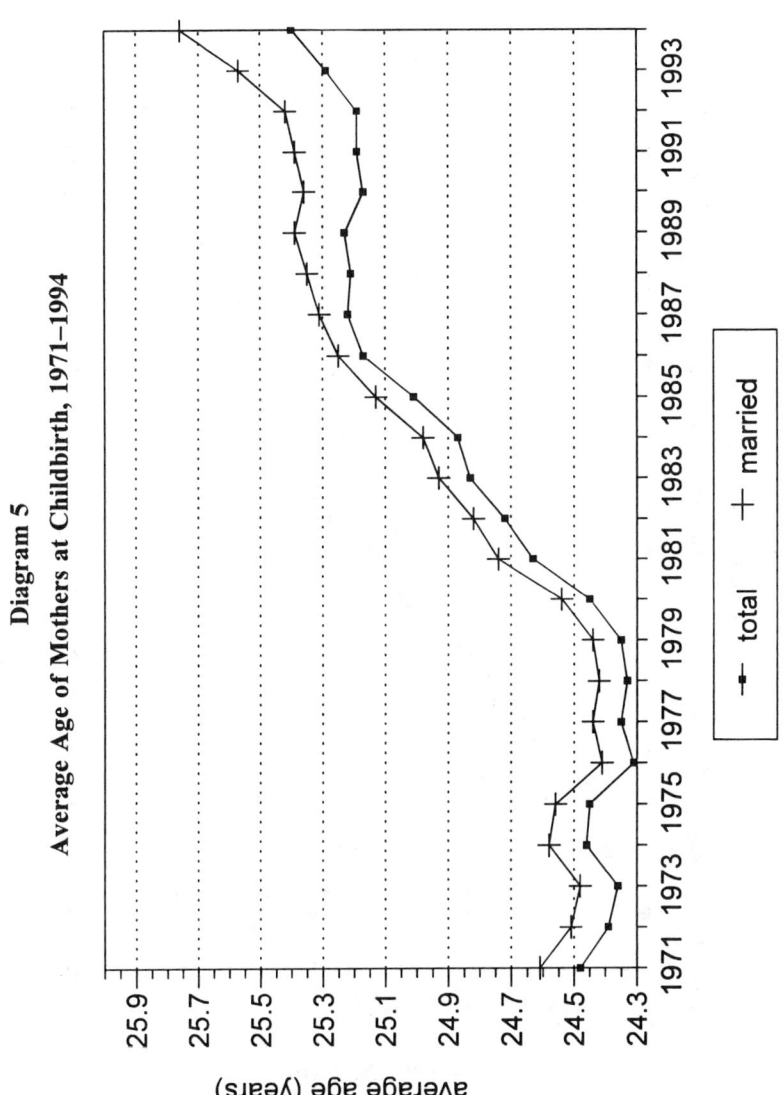

Diagram 5
Average Age of Mothers at Childbirth, 1971–1994

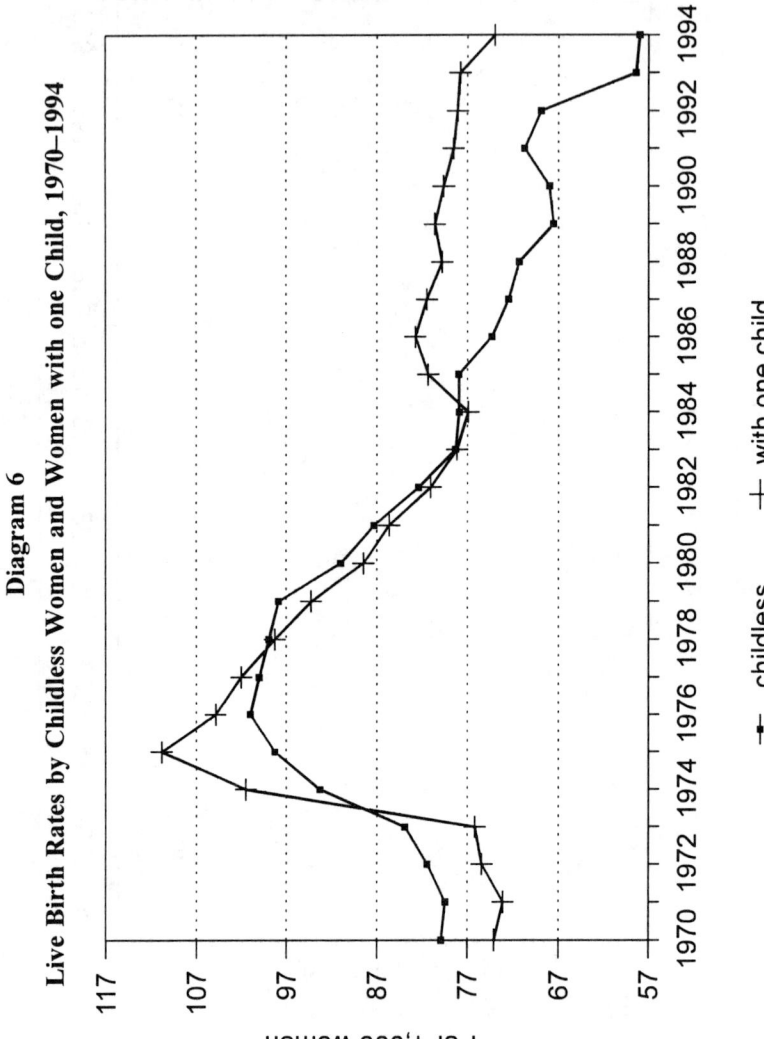

Diagram 6
Live Birth Rates by Childless Women and Women with one Child, 1970–1994

Birth Rates and Fertility in Hungary

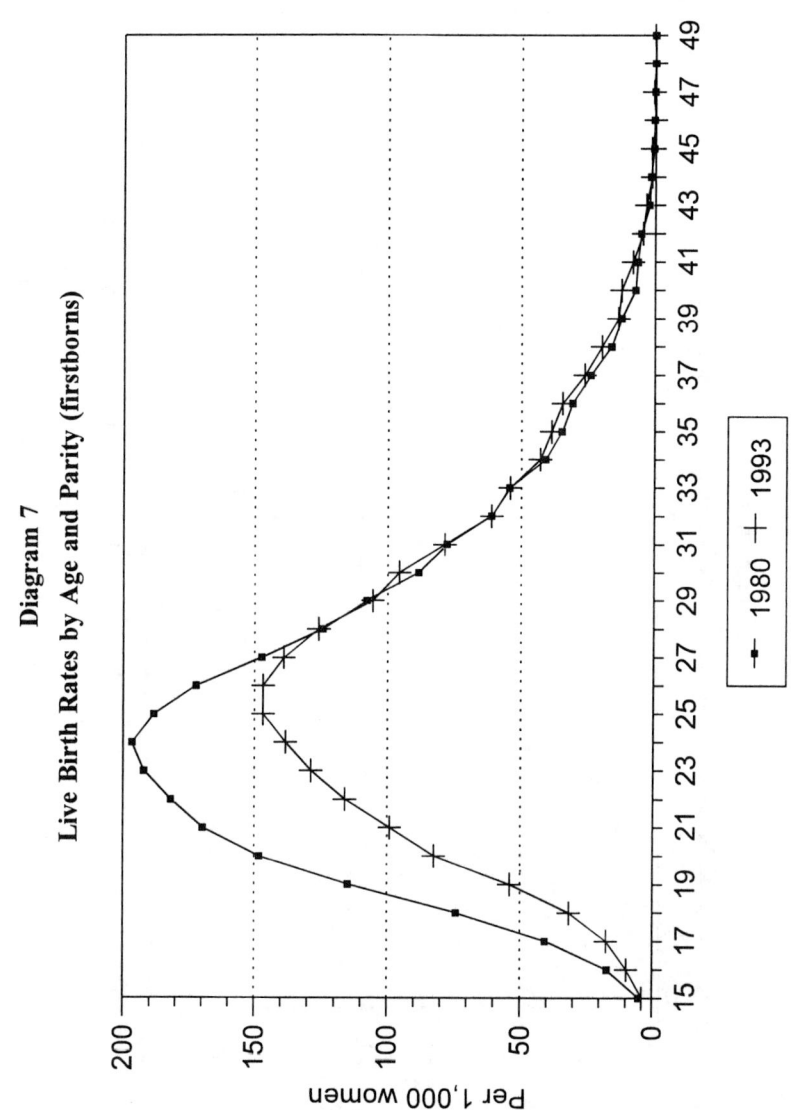

Diagram 7
Live Birth Rates by Age and Parity (firstborns)

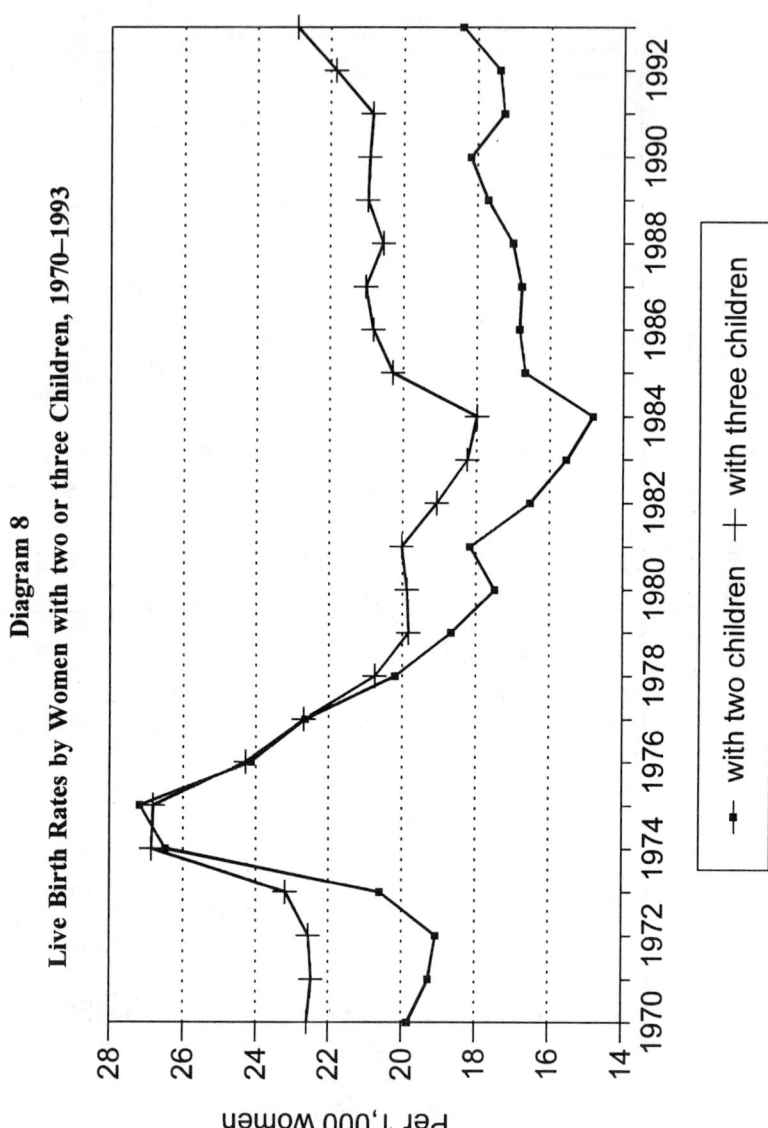

Diagram 8
Live Birth Rates by Women with two or three Children, 1970–1993

Birth Rates and Fertility in Hungary 81

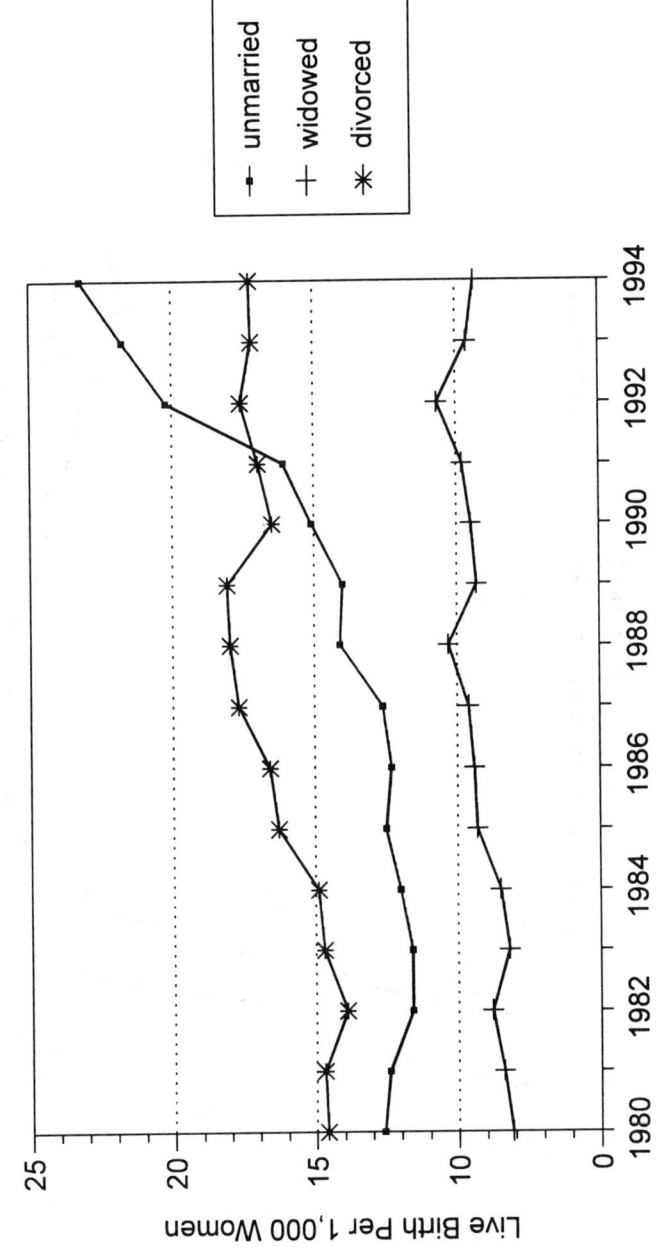

Diagram 9
Live Births Rates out of Wedlock by Marital Status

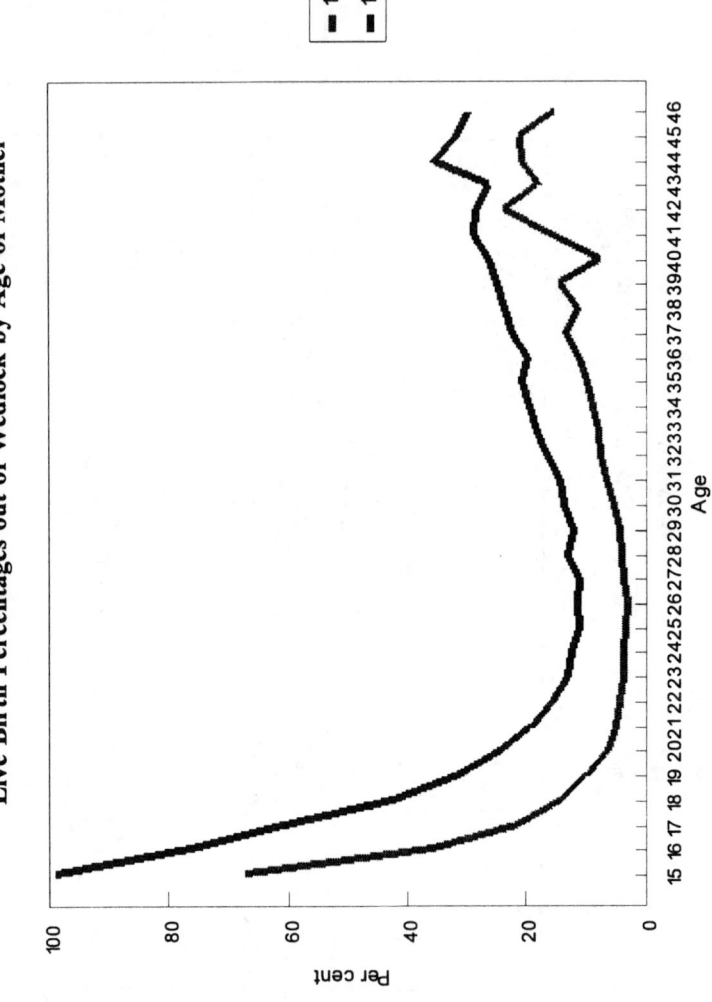

Diagram 10
Live Birth Percentages out of Wedlock by Age of Mother

Birth Rates and Fertility in Hungary 83

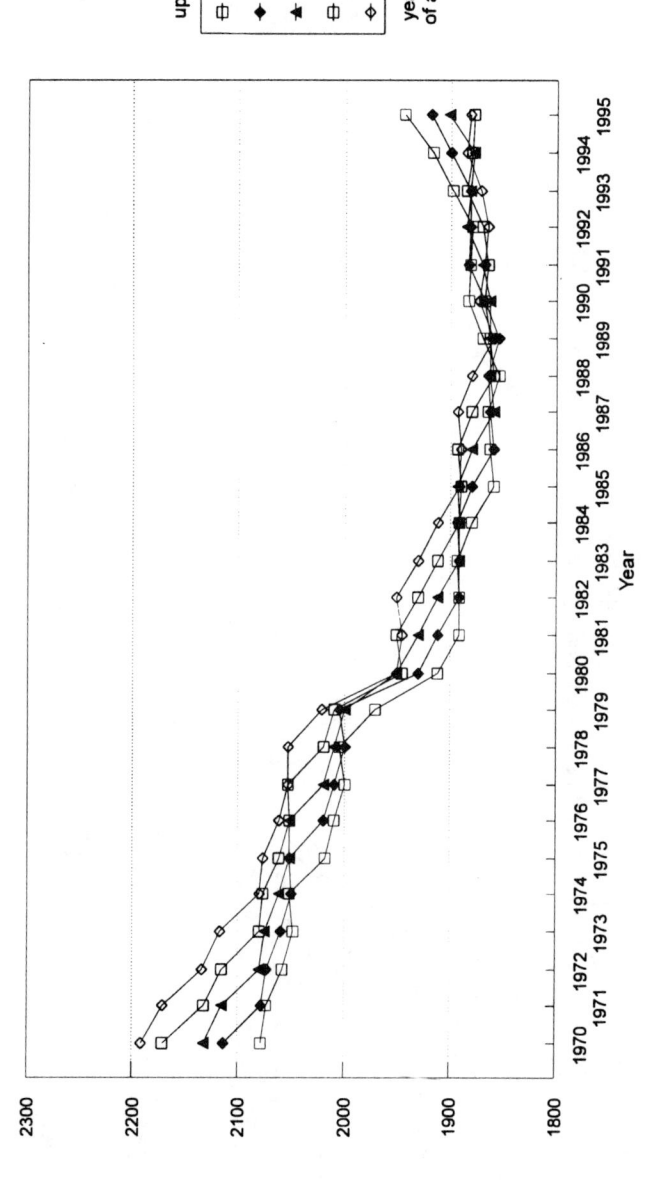

Diagram 11
Average Number of Descendants in the Female Age Group 45–49

Diagram 12
Average Number of Descendants in the Female Age Group 40–44

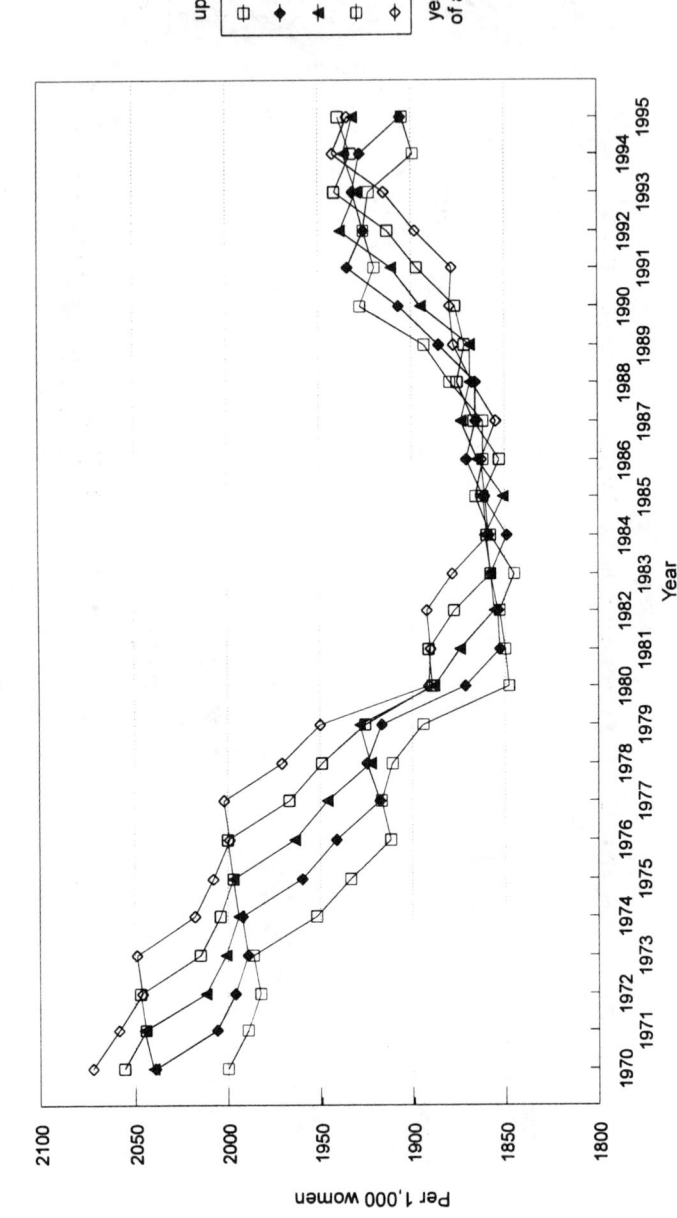

Table 1
Changes of Fertility by Age on Present-day Hungarian Territory Between 1900 and 1940

Average of years	Age group (year)							15–49 Total	Total fertility rate (per one women)
	under 17	17–19	20–24	25–29	30–39	40–49			
	15–19								
1900-1901	5.9	96.7	256.1	272.1	194.7	39.2		155.1	5.28
1910-1911	6.8	93.7	240.9	239.2	163.7	33.7		140.3	4.67
1920-1921		43.2	203.3	201.4	124.2	26.3		115.4	3.74
1930-1931		40.9	158.5	151.8	94.2	15.7		88.0	2.84
1940-1941		41.1	144.9	134.5	77.2	12.1		71.3	2.48
1900=100%									
1910-1911		–	94.1	87.9	84.1	86.0		90.5	88.4
1920-1921		–	79.4	74.0	63.8	67.1		74.4	70.9
1930-1931		–	61.9	55.8	48.4	40.1		56.7	53.8
1940-1941		–	56.6	49.4	39.7	30.9		46.0	47.0

Live birth per 1,000 women in an age group

Table 2
Live Birth By Age-Group of Mother

Year	Number of live births per 1,000 women in an age group							Total fertility rate (per one women)
	15–19	20–24	25–29	30–34	35–39	40–49	15–49	
1947	43.1	156.1	136.1	92.0	58.3	11.5	74.4	2.54
1948	45.6	162.8	140.9	88.0	56.3	11.2	76.1	2.58
1949	47.3	162.6	140.7	84.0	52.7	10.9	75.4	2.54
1950	51.5	170.7	141.0	89.2	50.8	10.6	77.4	2.62
1951	51.4	168.9	138.0	85.6	46.2	9.9	75.4	2.54
1952	48.1	165.3	135.4	85.3	43.5	9.2	73.5	2.48
1953	49.8	179.4	151.6	99.5	50.2	10.3	81.8	2.75
1954	52.0	195.6	164.7	107.7	54.3	10.2	88.3	2.97
1955	54.1	192.1	150.9	95.8	52.4	9.1	83.0	2.82
1956	54.4	183.9	135.8	83.2	47.0	6.5	75.9	2.59
1957	55.7	174.4	115.6	66.6	35.6	4.7	66.9	2.29
1958	54.4	169.9	110.5	60.8	31.0	3.6	63.6	2.18
1959	53.5	162.7	107.3	56.4	27.7	3.2	60.7	2.08
1960	52.5	159.2	105.6	52.9	25.0	3.6	58.9	2.02
1961	52.0	153.9	100.7	50.2	23.0	3.7	56.6	1.94
1962	46.4	143.1	94.8	47.0	20.3	3.7	52.5	1.79
1963	43.3	143.9	100.7	48.4	21.1	3.6	53.4	1.82
1964	41.8	145.1	101.3	48.0	19.1	3.4	53.2	1.80
1965	41.9	147.9	100.6	47.8	18.2	3.0	53.2	1.81
1966	45.8	152.7	104.8	48.5	18.8	2.9	54.5	1.88
1967	50.4	160.9	112.5	53.0	19.7	2.7	57.7	2.01
1968	52.2	164.5	116.3	54.4	19.7	2.4	58.7	2.06
1969	53.7	162.3	114.3	53.7	19.5	2.4	58.1	2.04
1970	50.0	159.3	110.3	51.4	18.4	2.2	56.6	1.97

Birth Rates and Fertility in Hungary 87

Table 2
(continued)
Live Birth By Age-Group of Mother

Year	Number of live births per 1,000 women in an age group							Total fertility rate (per one women)
	15–19	20–24	25–29	30–34	35–39	40–49	15–49	
1971	50.3	157.7	103.8	49.8	17.9	2.1	55.9	1.92
1972	53.5	157.4	105.2	47.8	17.4	2.2	56.9	1.93
1973	57.5	157.0	105.1	48.1	17.9	2.0	58.2	1.95
1974	67.1	180.5	128.6	59.9	20.0	2.3	69.6	2.30
1975	72.1	183.5	133.8	62.0	20.2	2.2	72.8	2.38
1976	74.5	178.1	121.9	54.2	18.2	1.9	69.9	2.26
1977	73.8	172.7	114.6	51.1	17.3	2.0	67.3	2.17
1978	73.5	169.4	101.3	47.3	15.6	1.7	64.1	2.08
1979	72.9	166.0	104.2	42.6	14.8	1.6	61.5	2.02
1980	68.0	158.6	100.0	40.9	13.7	1.5	57.6	1.92
1981	62.4	155.9	100.6	40.6	12.8	1.5	55.7	1.88
1982	58.3	149.5	95.6	38.5	12.3	1.4	52.2	1.78
1983	53.3	145.4	94.2	37.6	12.0	1.3	49.8	1.73
1984	52.1	146.1	94.6	39.1	12.1	1.3	49.0	1.73
1985	51.5	152.5	102.9	43.0	13.6	1.4	50.9	1.83
1986	48.8	148.3	107.1	43.9	14.5	1.4	50.1	1.83
1987	47.1	145.4	107.1	44.3	14.7	1.5	49.2	1.81
1988	44.1	143.5	109.5	43.5	14.7	1.5	48.5	1.79
1989	40.8	141.8	110.7	43.9	15.3	1.5	47.9	1.78
1990	39.5	147.2	115.4	46.9	16.4	1.6	49.4	1.84
1991	38.1	146.1	117.4	49.7	16.8	1.8	49.7	1.86
1992	35.6	136.3	113.5	48.0	16.5	1.7	47.3	1.77
1993	34.0	123.2	112.3	48.1	16.1	1.9	45.3	1.69
1994	33.6	113.9	110.3	50.2	17.2	2.0	44.7	1.64
1995	31.4	103.9	108.6	49.8	16.9	1.9	43.4	1.57

Table 3
Family Planning Survey Results (women)

Survey	Number of children (percentage)					Total	Average number of children
	0	1	2	3	4-x		
1974 Marriage cohort							
Planned number of children at the time of marriage (1974)	1	6	72	19	2	100	2.17
Actual size of family (1990)	4	19	57	16	4	100	2.01
Ideal family size (1990)	1	7	62	28	2	100	2.22
1983 Marriage cohort							
Planned number of children at the time of marriage (1983)	0	10	74	14	2	100	2.07
Desired number of children six years after the wedding (1989)	1	18	63	15	3	100	2.00
1991 Marriage cohort							
Desired number of children in the first year of marriage (1991)	1	15	61	19	4	100	2.12
1993 cross-sectional survey (women of 18-41)							
Total number of desired children	2	15	61	17	5	100	2.10

Károly Miltényi

Recent Changes in the Economic Activity and Retirement Patterns in Hungary: Financial and Social Implications *

Introduction

Hungary's economy was characterized by high labor-force participation rates in the middle-aged groups until recently. This was accompanied by a retirement pattern that was rather exceptional by international standards. Out of 151 countries listed by an ILO paper,[1] Hungary had the lowest LFPRs for males over sixty; female LFPRs were also among the lowest in the age group over fifty-five.

The recent changes, particularly the increasing unemployment, obviously have their effects on these rather stable employment/retirement patterns. This paper explores the potential implications of the changes on the financing of the pension system, considering also the projected population size and structure and the concomitant changes of the support ratio.

Retirement in recent years and in the near future has affected and will affect mainly the cohorts starting their working career in their fifties or early sixties. After a rather stable career of thirty to thirty-five years, these cohorts are reacting, understandably, to the recent changes in a manner that had been conditioned during this period. Thus, their situation and career strategy is still influenced partly by the 1949–90 trends. It is for this reason that the first section of this paper, gives a short historic background which is to some extent more explanatory when studying the attitude, preferences, and behavior of the old labor-force than factors shaping the present situation.

---

* This paper is a somewhat modified and extended version of the author's paper submitted and discussed at the international conference on demographic processes and the socioeconomic transformation in Central and Eastern European countries (June 8–11 1995, Jachronka, Poland) for the Demographic Research Research Institute, Budapest.

I. Economic activity: 1949–90 Trends

There were two basic trend periods in the labor-force participation rates between 1949 and 1990 (Table 1). The first period (1949–75) was characterized by a rapid increase in the economically active population by employing cheap additional labor-force, mostly housewives. The increase of the employed population in this period was about 1 million, which was entirely due to the increase of the female employment (see Table 2): in the male LFPRs there was a slight decrease. The activity ratio reached the maximum (48,4%) and the dependency ratio the minimum (106) in 1975. In the second period (1976–90) the trend reversed: activity ratio decreased and dependency ratio increased more or less continuously and at an accelerating pace (see Figure 1).

The factors influencing these trends were partly of demographic and partly of social character. Around 1950, male participation rates were nearly 100% in the relevant age groups, with only disabled persons excluded (see Table 3). This was necessitated by the fact that after the nationalization of factories, workshops, more-than-one-flat houses, etc., and the incorporation of the private farms in to the agricultural cooperatives, practically no capital income remained. Rapid industrialization, on the other hand, created new job opportunities, especially in the urban areas.

Thus, labor-force participation rates could be increased only among women, where labor-force reserve was still significant (see Table 3). The increase in female participation was in harmony which the declared social and labor policy. However, it was also necessitated by the low income level; a family with only one earner had to face serious financial difficulties. Consequently, the family model of two earners and one or two children became general in the period 1950–70.

However, in 1968 a significant change took place in the female employment policy. Motivated mainly by population policy considerations, a child-care system was introduced. This mades it possible for the mother to remain at home until the child reached the age of three. She was on leave without payment and received a social security benefit/payment. Since the introduction of the system—depending on the changes in the number of births—200–290 thousand women were on child-care leave, i.e., 4–5% of the total and 8–12 % of the female labor-force. Paradoxically, this institution has increased the legal labor-force participation rates of the young women, since it made the family-oriented women interested in joining the labor force as early as possible. (Only women employed for a given period were entitled to this benefit; this was changed only recently.)

The other significant institutional factor in the change of the labor-force trends was the extension of the social security system, which gradually entitled almost all strata of the population to old-age pension, the system of which was also improved gradually in 1960–75. Consequently, after acquiring the required work period, most of the workers and employees reaching the rather low age limit for old-age pension (sixty for males and fifty-five for females) opted for retirement. Thus, LFPRs in old-age groups fell dramatically in 1960–90, and after 1976 the labor force started decreasing.

II. Retirement Patterns: Background and Recent Changes

Due to processes outlined above, both the number of pensioners and especially the pension expenditures have increased rapidly (see Table 4 and Figure 2). Whereas in 1960, 8% of the population were pensioners and social security expenditure, determined mainly by the pension costs, constituted 4% of the GDP, the same proportions were 28% and about 24% respectively in 1994. Considering the projected population size and structure (*ceteris paribus*) expenditure would explode after 2000.

Although life expectancy is unfavorably short in Hungary (64.5 for males and 73.8 for females), due to the low retirement age life expectancy in retirement is among the longest by international standards, i.e., 14.5 years for males and 23.1 for females (see Table 5). Moreover, if we consider early and disability retirement, the duration of pensioner years is still longer. It has to be emphasized that out of one hundred males starting their economic activity, only forty-five reach age sixty in this status, thirty-five retire (early or disability retirement), and twenty die before reaching this age (see Table 6). For females the situation is different; due to their lower mortality and lower (fifty-five) retirement age, out of one hundred economically active women, seventy-four remain active until fifty-five.[2]

Immediate /early/ disability retirement was reinforced by the recent economic changes inducing increased structural unemployment. The yearly number of retirements was around 130,000 during 1985–88; it increased to 170,000–210,000 in the 1990–94. This increase was entirely due the increase of the early and disability retirement (Table 9), encouraged also by the government and intended to avoid unemployment. It is for this reason that unemployment affects mainly the young age groups; over-fifty unemployment is rather exceptional (see Table 7). The high unemployment rates of the young age groups are the main factors influencing the drastic fall of nuptiality and fertility.

The higher male unemployment, rate especially in the young age groups, also has an institutional explanation; substantial part of the female employed (11% in general and more than 30% in the age group twenty to twenty-nine) were in 1990 on child-care leave, during which dismissal and the consequent unemployment is legally prohibited.

Thus, recent economic and social changes e.g., increasing unemployment, income-generating activities in the economy private /second/ informal have reinforced, especially among the manual workers, the inclination toward retirement. This is also influenced by the unfavorable health conditions. Hungarian mortality is among the highest in Europe, especially male mortality. Life expectancy of adult males has decreased in the last thirty-three years: for males aged 40 it was 32.2 in 1960 and 27.9 in 1993, for males aged 50 it was 23.3 in 1960 and 20.6 in 1993. The unfavorable morbidity situation is reflected also by the 1986 health sample-survey data, covering 37,500 persons, or 0,4% of the population (see Table 8). Even if we make allowances to the partly subjective nature of the complaints (especially those coming from females), at least one third of the population aged forty to fifty-nine has serious health problems, and only 22% in this age group considers him/herself healthy without complaints.[3] This, undoubtedly, may influence career decisions. So does the high male mortality, experienced and felt by the general public, even without knowing technical or statistical details.

III. The Financing System of the Pensions: Implications of the Aging

As in most European countries, pensions are financed in Hungary from the current social security contributions, i.e., in a pay-as-you-go system. This situation will prevail in the foreseeable future.

The relationship between population aging and old-age pension financing in the pay-as-you-go system can be illustrated in a simple model.[4] All others being equal and assuming a fixed replacement rate, i.e., a fixed ratio between average pension and average earnings the social security tax rate is determined by the support ratio, i.e., the number of active workers per pensioner. Obviously, the support ratio is inversely proportional to the required tax rate.[5]

Social security contributions are paid in Hungary by the employer (44%) and the employee (10%), i.e., together 54% of the earnings is the source of the social security benefits. Out of total social security expenditure, 56% is the pension expenditure (1990). This created a more or less balanced situation, as a 30% pension contribution after earnings combined with the 1990 1.8 support ratio (4.5 million active earners and 2.5 million pensioners) ensured a 54% replacement rate, i.e., it made possible that even in an inflation period average pension should be about 55 of the average earnings, which has been the case in Hungary in the last five to six years.

Population projections for 1996–2010 indicate that the slow decrease and the concomitant aging of the population will continue in Hungary. From 2000 there will be a rapid increase in age groups over present retirement age, i.e., sixty for males and fifty-five for females. Simultaneously, labor-force age groups will decrease, and considering also possible unemployment, support ratio may go down from the 1990 1.8 to 1.2–1.3.

There are three possible options to maintain the present almost-balanced situation.

1. To increase the percentage of the social security/pension contribution to maintain the 50–55% replacement rate and to compensate for the decrease in the support ratio. This, together with the expected increase of the health insurance cost would require a 60–65% social security contribution against the present 54%, already very high by international standards.

2. Theoretically, replacement ratio, i.e., the proportion of pensions to active earnings, could be reduced, too. However, considering the relative poor and vulnerable position of the pensioners, no deliberate and explicit policy can be formulated on this basis. On the other hand, inflation without a general and systematic pension-adjustment system works in this direction. (The present adjustment system operates on an ad hoc basis in Hungary and affects mostly the lower than average pensions. Recently, there is a tendency to increase the pensions by the same rate as active salaries are increased in the formal sector.)

3. To maintain the present support ratio by increasing retirement age. Long-term policies are contemplated to increase retirement age for both sexes to sixty-two or sixty-three, starting a gradual increase process for

the females in 1995 and completing the transition around 2005. The parliament has taken a general decision in this respect, however the practical steps concerning the implementation of the pension reform are still being disputed.

Apart from the potential unemployment-increase effect, the main argument against the changes in the retirement age is that they may induce further increases in the disability retirement. Western European and North American countries with higher (sixty-five to sixty-seven years) retirement ages experience, undoubtedly, similar tendencies, i.e., a significant part of the economically active persons around and over sixty opt for disability/early retirement.

In my opinion this argument has only partial validity. It is right in emphasizing that retirement-age increases must be followed by radical changes in the present overliberal disability retirement system and, of course, by the gradual improvement in the health status of relevant age groups. However, it seems to me that official retirement age per see has significant social and psychological implications. It strongly influences the attitude and the whole life strategy of the population. Irrespective of the various professional or scientific definitions, public opinion considers retirement age as identical with old age, and expects that members of society should generally behave in conformity with the rules determined by this age, in their economic activity as well. Thus, it is possible, or even probable, that a change in the retirement age will affect the attitude and behavior not only in the relevant age groups but also in the whole population.

The dramatic decrease of economic activity in the formal sector is one of the most serious problems of our present and future economy. The number and proportion of the inactive households, i.e., households without any active earner, has doubled between 1970 and 1990, and will increase further by 2010. Thus, provided the trend continues, in 2010, 44% of the total households will consist only of economically inactive members (see Table 10 and Figure 3).[6]

The main factor in this unfavorable process is, undoubtedly, aging. However, the institutional background and the legal practice also have a significant impact. The over-liberal practice of disability retirement, unemployment benefits, child care allowance, etc, enables a large proportion of families to apply special strategies for their survival. They receive some fixed, modest social benefits. In addition to this, they are engaged in various activities in the second/black economy, which produces about 20–30% of the GDP. However, the spread of this practice is, of course, self-defeat-ing. The economic activity of the black economy may contribute to the GDP, but by avoiding the payment of social security contributions and taxes it undermines the funding of the various social-policy benefits covered exactly by these contributions/taxes.

Thus, basic reforms in the pension system, and generally in the system of social-policy benefits, financed from the social insurance and the state budget, are inevitable. Simultaneously, the increase of employment in the formal/legal economy is a precondition for a balanced secondary income distribution, i.e., pension, family allowance, and other benefits. This is one of the reasons for the strict measures introduced by the government in March 1995, cutting down expenses in the social-policy. However, concentrating social policy benefits on the poorest strata of the population and excluding middle and upper strata from the various family/child benefits may induce strong differential fertility, increasing the proportion of the poorest and least-educated strata.

IV. Long-Term Economic Implications: the Problems of Population/Family/Social Policies in Perspectives.

There is a somewhat general consensus among demographers—supported by various optimum calculations—that the optimal population is around the stationary population, i.e., a population without considerable increase or decrease, and consequently, with more-or-less constant and proportional age structure. Increasing population implies a strong child-dependency burden and necessitates demographic investments of considerable magnitude. This—as in the case of the developing countries—requires 10–15% of the GDP just to maintain the status quo.

Whereas the unfavorable consequences of a rapid population growth are well known for a long time, the economic implications of the population decrease and the concomitant aging, i.e., increasing old-age dependency, have been studied only in the last ten to fifteen years. This was due to the fact that the pension schemes introduced in Europe in the fifties or sixties had to be revised after 1980, as the aging of the population made their continuation impossible.

For the next ten to fifteen years, the three possible reforms outlined above may solve the problems of financing the system. However, it is obvious that long-term consequences, i.e., the support ratio after 2010, will be determined by present and future fertility. Thus, if we want to avoid the further dramatic deterioration of this problem, a strong and efficient pro-natalist population policy is needed.

This has to be emphasized vis-à-vis the declarations of some liberal economists. According to these declarations, any interference in the "spontaneous"* population processes is as deplorable as the interference in the supply-demand regulated free-market of goods and services. Implicitly, these views assume an "invisible hand" also in the human population, influencing the "spontaneous" population processes toward an optimum, i.e., stationary, population.[7]

Unfortunately, empirical evidence both in the world population and in the national populations (including Hungary) indicate that these views are completely unfounded. Nowhere in the world do population processes move—or have ever moved—automatically toward a stationary population. In other words, fertility decisions and behavior do take place on the micro-society/family level, and their outcome is totally different from the fertility that can be considered optimal for the macro-society. The assumption that in the long term a "balanced population" will develop turned out to be wrong both for the developing and for the developed countries.

Summary and Conclusions

Hungary's economy was characterized by high labor-force participation rates in the middle-aged groups until recently. This was accompanied by a retirement pattern that was rather exceptional by international standards. Out of 152 countries listed by an ILO paper, Hungary had the lowest LFPRs for males over sixty; female LFPRs were also among the lowest in the age group over fifty-five.

From 1989 to 1990, due to economic transformation and the concomitant structural unemployment, economic activity, especially in the formal sector, decreased sharply. As the young age groups have the highest unemployment rates, this reduced dramatically nuptiality and fertility. At the same time, inclination for early/disability retirement has increased. The

---

* It has to be noted that in any civilized society "spontaneous population" is a fiction. Socioeconomic factors, cultural background, and traditions do basically influence the fertility attitude and behavior of couples. In addition to this, legislation, institutional background, and government policies also have strong effects on fertility, irrespective of whether these effects are intended or not.

percentage of the pensioners rapidly increased, from 8% in 1960 to 28,3 in 1994. Social security expenditure was about 4% of the GDP in 1960, and 24% in 1994.

Due to the natural decrease of the population and the concomitant rapid aging, the present system of pension financing cannot be continued. The support ratio, i.e., ratio of active earners to pensioners, of 1,8 in 1990 may go down to 1,2–1,3 after 2000, thus even the high social security contribution rate (54%) cannot ensure the present replacement ratio, i.e., that the pensions should be about 50–60% of the active earnings.

The increase of the informal/black economy, producing now about 20–30% of the GDP, is contributing to this crisis. The widespread practice of families to utilize the over-liberal system of disability retirement, unemployment benefits, child-care allowance, and, at the same time having additional income from the black economy, undermines the present social-policy system. As the black economy does not pay social security contributions and taxes, this strategy applied by some part of the inactive households, i.e., a third 1/3 of the total households, cannot maintain the present social-policy system. However, concentrating social-policy benefits on the poorest strata of the population and excluding middle and upper strata from the various family/child benefits may induce strong differential fertility, increasing the proportion of the poorest and least-educated strata.

The assumption that "spontaneous" population processes may create, or, at least, approach the optimal population is wrong and refuted by all empirical evidence, both on a world-wide level and in Europe, including Hungary. It is also wrong to assume that a population decrease is unfavorable only from a specific emotional/ideological point of view. The present and projected population decrease of Hungary with its concomitant aging has, and will have, serious economic implications. Thus, pro-natalist policies are necessitated by the purely pragmatic aspects of economic rationality.

Figure 1
Population by Economic Activity

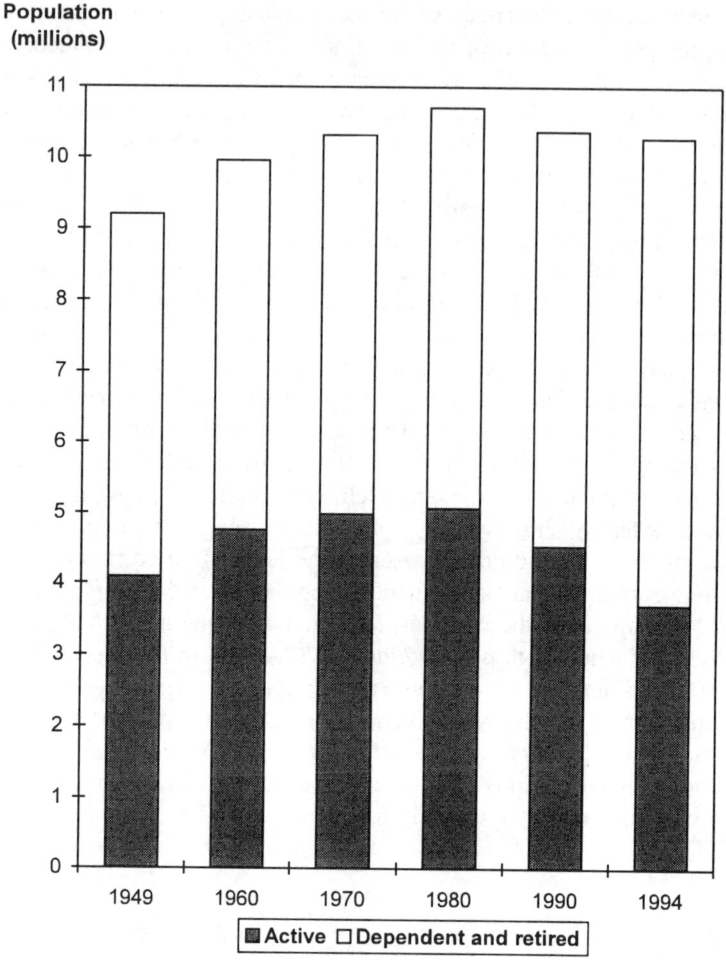

Figure 2
Basic Data of Pensioners

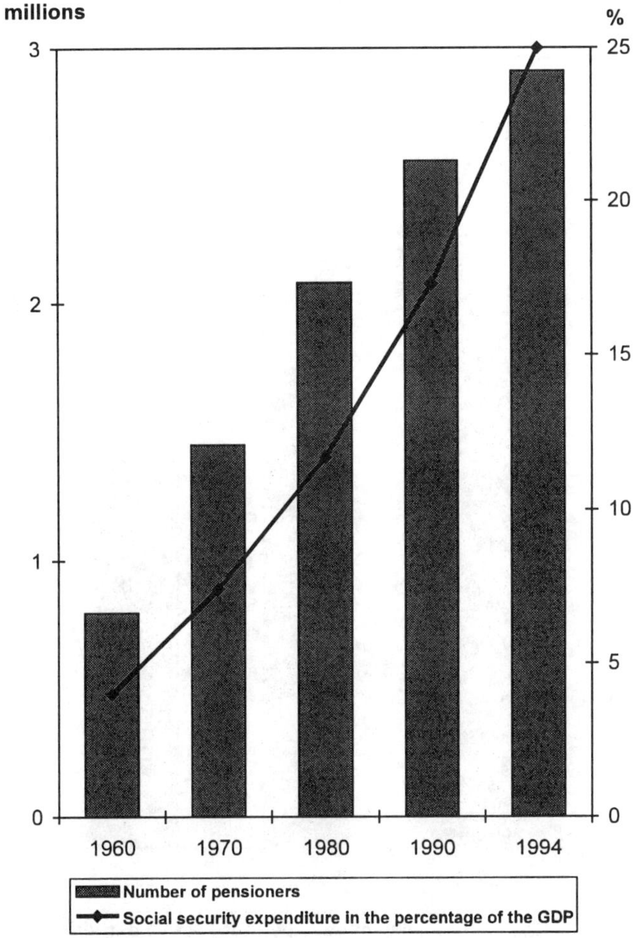

- Number of pensioners
- Social security expenditure in the percentage of the GDP

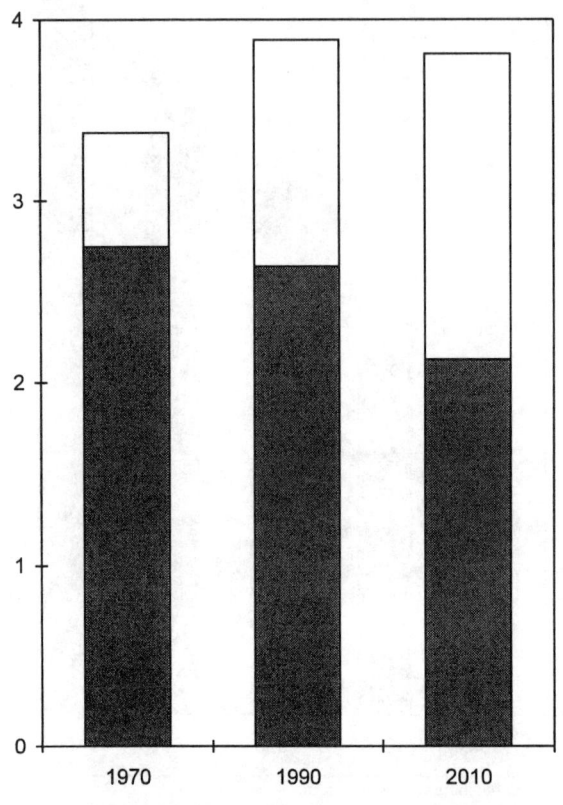

Figure 3
Number of Households by Economic Activity

Table 1
Population by Economic Activity

Year	Population	Economically active	Dependent and retired	Activity ratio[1]	Dependency ratio[2]
	1000				
1949	9,204.8	4,084.9	5,119.9	44.4	125
1960	9,961.0	4,759.6	5,201.4	47.8	109
1970	10,322.1	4,988.7	5,333.4	48.3	107
1975	10,500.9	5,085.5	5,415.4	48.4	106
1980	10,709.5	5,068.8	5,640.7	47.3	111
1985	10,657.4	4,912.9	5,744.5	46.1	117
1990	10,374.8	4,527.2	5,847.6	43.6	129
1991	10,354.8	4,668.7	5,686.1	45.1	122
1992	10,337.2	4,241.8	6,095.4	41.0	144
1993	10,310.2	3,866.9	6,443.3	37.5	167
1994	10,277.0	3,700.7	6,576.3	36.0	178

1 Percentage of economically active population to the total population.
2 Number of dependent and retired population per 100 economically active population.

Table 2
Female Population by Economic Activity

Year	Population 1000	Economically active 1000	Percentage of females among the economically active population	Female activity ratio
1949	4,781.4	1,192.9	29.2	25.0
1960	5,157.0	1,691.1	35.5	32.8
1970	5,318.4	2,055.2	41.2	38.6
1975	5,411.2	2,235.4	44.0	41.3
1980	5,520.8	2,202.0	43.4	39.9
1985	5,508.1	2,247.2	45.7	40.8
1990	5,389.9	2,013.5	44.5	37.4
1991	5,382.6	2,125.8	45.5	39.5
1992	5,376.7	1,965.3	46.3	36.6
1993	5,366.8	1,776.6	45.9	33.1
1994	5,354.0	1,743.6	47.1	32.6

Recent Changes in the Economic Activity 103

Table 3
Labor force Participation Rates (by Percentage)
by Sex and Age, 1949–94

Age group	1949	1960	1970	1980	1984	1990	1992	1993	1994
Males									
14	2.5	17.7	3.7	1.5	1.5	0.6	–	–	–
15–19	63.6	57.2	45.8	45.5	43.5	34.7	24.4	22.3	21.4
20–24	92.6	94.5	91.5	91.9	89.9	85.7	81.1	79.3	75.2
25–29	97.0	98.8	98.5	98.2	98.3	93.9	92.7	91.7	91.6
30–34	98.4	99.0	98.6	98.4	98.3	94.7	93.5	92.0	90.7
35–39	97.8	98.8	98.1	97.8	97.7	94.1			
40–44	97.5	98.5	97.2	96.0	96.2	92.5			
45–49	96.1	97.9	95.4	92.9	93.0	89.2	86.4	83.8	82.8
50–54	93.6	96.7	91.8	86.2	86.0	80.5			
55–59	88.4	93.4	84.4	72.2	70.1	61.0	52.3	48.1	44.2
60–64	80.7	69.6	43.7	13.2	5.5	3.8			
65–69	72.8	65.5	24.6	5.3	1.8	1.7	13.9	11.4	9.2
70–X	53.8	51.3	10.7	3.0	1.0	0.6			
Total	65.4	63.9	58.6	55.3	53.3	49.7	66.9	64.3	62.6
Females									
14	2.3	13.4	10.5	3.6	3.5	2.4	–	–	–
15–19	52.7	48.1	49.0	40.4	37.9	31.4	23.0	21.4	19.2
20–24	45.0	55.1	66.1	59.9	62.7	59.1	78.6	75.5	72.4
25–29	33.8	48.7	65.3	69.8	71.4	62.1	89.5	87.8	86.9
30–34	30.9	49.1	68.7	81.1	83.0	76.9	89.9	87.2	85.2
35–39	28.6	50.7	71.0	84.9	88.9	85.4			
40–44	27.6	51.8	69.4	83.1	87.4	87.6			
45–49	27.0	49.7	64.0	77.5	82.7	82.1	78.2	76.0	73.2
50–54	25.4	46.3	56.6	67.4	70.7	66.8			
55–59	26.7	30.6	29.2	18.8	9.5	5.6	19.7	17.3	14.0
60–64	25.8	26.1	17.1	8.7	3.7	1.6			
65–69	22.7	24.1	9.5	5.1	1.6	0.7	8.4	6.5	5.2
70–X	17.1	18.6	3.5	2.1	0.3	0.2			
Total	25.0	32.8	38.6	39.9	39.5	36.9	57.1	54.5	52.0

Table 4
Basic Data of Pensions

	1960	1970	1980	1990	1992	1993	1994
Number of pensioners (1,000)	796	1,453	2,082	2,556	2,751	2,840	2910
Percentage of the total population	8.0	14.1	19.4	24.7	26.6	27.6	28.3
Average monthly pension (Ft)	486	256	2,267	6,683	9,747	1,1503	13,553
Percentage of the average earning	31.4	35.9	55.1	66.1	62.4	62.5	
Social security expenditure in the percentage of the GDP	4.0	7.4	11.7	17.3	19.9	23.4	

Table 5
Life Expectancy and Retirement Age in Some Countries
(Latest Available Data)

Number	Country	Life expectancy		Retirement age		Life expectancy in retirement	
		male	female	male	female	male	female
1.	Sweden	73.0	79.1	67	67	13.6	17.2
2.	Norway	72.5	79.2	67	67	13.3	16.7
3.	Holland	72.4	79.2	65	65	14.3	19.1
4.	Switzerland	72.0	78.7	65	62	15.0	22.7
5.	Denmark	71.1	77.2	67	62	13.1	19.4
6.	Great-Britain	70.4	76.6	65	60	13.6	21.3
7.	Spain	70.4	76.2	60	55	18.4	26.5
8.	Canada	70.1	77.4	66	66	12.1	17.4
9.	Greece	70.1	73.6	65	60	13.1	19.6
10.	USA	70.0	77.8	65	65	14.5	18.4
11.	France	70.0	78.2	60	60	18.4	23.6
12.	FRG	69.9	76.6	63	63	15.2	19.3
13.	Italy	69.7	76.9	60	55	17.9	26.9
14.	Finland	69.5	77.8	66	66	12.8	16.7
15.	GDR	69.5	75.4	65	60	12.5	19.4
16.	Austria	69.0	76.2	65	60	13.1	20.9
17.	Ireland	68.8	73.5	66	66	10.8	15.5
18.	Bulgaria	68.7	73.9	60	55	17.8	23.1
19.	Belgium	68.6	75.1	65	60	13.0	20.9
20.	Yugoslavia	67.7	73.2	60	55	15.7	24.2
21.	Romania	67.4	72.2	60	55	15.4	23.2
22.	Poland	67.1	75.2	65	60	12.9	19.9
23.	Czechoslovakia	67.0	74.3	60	60	14.9	19.3
24.	Hungary	64.5	73.8	60	55	14.5	23.1
25.	Portugal	65.1	72.9	65	60	18.4	26.5
26.	USSR	64.0	74.0	60	55	12.0	25.0

Table 6
Table of Mortality and Retirement of Economically Active Persons by Sex, 1984

Age	N[1]	Age group	In the age group indicated			Total leaving labor force
			deceased	retired		
				old age	disability	
Males						
14	100,000	14–29	1,890	–	1,127	3,017
30	96,983	30–39	2,768	–	2,449	5,217
40	91,766	40–49	5,616	–	7,234	12,850
50	78,916	50–54	3,592	1,031	7,666	12,289
55	66,627	55–59	5,482	3,797	11,733	21,012
Subtotal		14–59	19,348	4,828	30,209	54,385
60	45,615	60–64	2,363	36,338	5,820	44,521
65	1,094	65–69	125	717	81	923
70	171	70–X	66	97	8	171
Females						
14	100,000	14–29	520	–	833	1,353
30	98,647	30–39	937	–	2,694	3,631
40	95,016	40–49	1,706	321	6,921	8,948
50	86,068	50–54	1,695	1,340	8,948	11,983
Subtotal		14–54	4,858	1,661	19,396	25,915
55	74,085	53–59	1,302	59,644	3,985	64,931
60	9,154	60–64	257	5,698	82	6,037
65	3,117	65–69	102	2,000	20	2,122
70	995	70–X	363	627	5	995

1 N = Number of persons surviving and remaining in the labor force.

Recent Changes in the Economic Activity 107

Table 7

Unemployment Rate and Distribution by Sex and Age, 1992–94

| Age group | Unemployed per 100 economically active ||||||||| Percentage distribution of the unemployed |||||||||
| | total ||| male ||| female ||| total ||| male ||| female |||
	1992	1993	1994	1992	1993	1994	1992	1993	1994	1992	1993	1994	1992	1993	1994	1992	1993	1994
15–19	26.1	31.7	28.0	28.2	35.0	30.9	23.7	28.1	24.6	11.3	11.3	10.7	10.8	10.7	10.4	12.1	12.2	11.2
20–24	12.4	15.0	14.3	16.2	19.5	18.1	8.4	10.1	10.0	15.7	16.0	17.0	17.6	17.8	18.7	12.7	13.1	14.4
25–29	9.9	11.0	9.6	11.8	12.8	11.6	7.9	9.1	7.5	12.4	11.6	11.8	12.7	11.5	12.0	11.9	11.7	11.4
30–39	8.8	10.5	9.5	10.1	12.0	10.8	7.5	8.8	8.2	28.9	28.0	27.9	28.1	27.2	26.6	30.2	29.2	29.9
40–54	7.4	8.9	7.9	8.2	10.7	9.3	6.4	7.1	6.5	27.5	28.0	28.1	26.2	28.0	27.9	29.4	28.0	28.5
55–59	6.1	8.7	6.3	6.3	8.6	7.1	5.5	9.0	4.4	2.8	3.0	2.2	3.4	3.4	2.9	2.0	2.4	1.1
60–74	4.2	9.3	10.2	3.9	6.5	6.9	4.6	12.7	14.4	1.4	2.2	2.3	1.2	1.4	1.4	1.7	3.4	3.6
Total	9.3	11.3	10.2	10.7	13.2	11.8	7.8	9.2	8.4	100.0	100.0	100.0	100.0	100.0	100.0	100.0	100.0	100.0

Table 8
Percentage Distribution of the Population by Health Status and Age, 1986

	Total	out of which			
		0–14	15–39	40–59	60–X
Healthy, without complains	43	82	50	22	11
Deficient/disabled	5	1	3	5	11
Chronic illness(es)	18	3	9	28	42
Complaints only	35	14	38	45	36
	100	100	100	100	100

Table 9
New Retirements, 1985–94 (thousands)

	1985–88 (average)	1990	1991	1992	1993	1994
New retirements: total	130	189	212	199	185	170
from old age pension	–	99	109	117	103	90
disability	–	61	66	64	63	62
early	–	–	–	46	43	41

Table 10
Number of Households by Structure and Economic Activity, 1970–2010 (Thousands)

Household structure	Total			With active earner(s)							No active earner[1]		
				Together			Out of which						
							1			2 and more			
	1970	1990	2010	1970	1990	2010	1970	1990	2010	1970	1990	2010	1970	1990	2010
	(1) = (2) + (5)			(2) = (3) + (4)			(3)			(4)			(5)		
Total	3378	3890	3812	2746	2641	2125	1192	1202	.	1555	1439	.	631	1249	1687
Couple with or without child(ren)	2269	2282	1955	2003	1782	1355	706	536	.	1297	1245	.	266	500	600
One parent family	247	411	415	215	352	345	134	255	.	81	97	.	32	59	70
Single person households	590	946	1136	286	321	286	286	321	.	–	–	.	304	625	850
Other households	272	251	306	243	186	139	67	88	.	177	97	.	29	65	167

1 Only inactive earner (pensioner), dependent or unemployed.

Notes

1. Anker, R., and Clark, R. L.: "Labor-Force Participation Rates of Older Persons: an International Comparison." ILO paper, July 1989.
2. Klinger, A.:"Hungarian Retirement and Life Table." *Statistical Review*, Budapest, January 1988.
3. Central Statistical Office: The Health Status of the Population. Budapest, 1989.
4. Halter, W.A., and Hemming, R. "The Impact of Demographic Change on Social Security Financing." IMF staff papers, September 1987.
5. Weawer, C. L.: "Social Security in Aging Societies." *Population and Development Review*, (A supplement to volume 12), 1986.
6. Miltényi, K.: "A család és háztartásszerkezet várható alakulásának néhány gazdasági összefüggése." (Some economic connections of the expected family household structure changes, Presentation at the 1994. Conference on Family.) *Demográfia* 1994. 3–4.
7. Miltényi, K.: "Megjegyzések Bródy András: "Kis magyar demográfiá"-jához." (Notes on András Bródy's Little Hungarian Demography) *Kritika*, No 9, 1994.

Péter Józan

Changes in Mortality in Hungary between 1980 and 1994

The deterioration of mortality conditions in Hungary during the last few decades assumed epidemic dimensions by the early 1990s. This is the only way to describe the situation, since the mortality of the adult male population evokes the mortality rates of the 1920s and 1930s. This means that the mortality rate for the male age group of 35–64 in Hungary is among the worst in the world. In other words, men at the peak of their lives die to a higher degree than in most countries furnishing reliable data. As regards the extremely high level of mortality and the resulting bad probability of surviving, we have really fallen to the level of the "third world". The premature death of a person is tragic enough in itself, but it is still more so if we consider the economic losses due to the broken professional careers.

Besides, death strikes more often on the lower levels of society than it does on the higher ones, which calls attention to inequality in face of death.

Although there may really be troubles in the health services, the increasing frequency of deaths is rather due to illnesses, the prevention of which is not primarily the task of the health service. People have to understand that the decrease in the probability of surviving is primarily due to the unhealthy lifestyle of large masses of the adult male pupulation. The rise of mortality due to smoking and alcoholism can partly be seen as an epidemic. The question is whether the society is capable of curing such an epidemic.

The critical situation today is mostly a heritage of the past, but putting an end to it falls within the scope of the political processes of the country. Without a definite political will in this direction, the crisis cannot be overcome. If massive health improvement of the population is not a political priority, health conditions and mortality will most probably further deteriorate.

Mortality in Hungary is about sixty-five per cent higher than in Austria. This exceptionally high mortality, the worst among industrial countries with well developed health services, is one of the causes of the natural decrease of the number of the population. So the epidemiological crisis leads to a demographic one.

Table 1

Male Life Expectancy in Hungary and Austria at the Age of Thirty and Forty-five between 1920–21 and 1993

	1920–1921	1930–1931	1966	1993
Remaining life-time to be expected at the age of 30 (in years)				
Hungary	34.93	37.09	41.99[b]	36.50
Austria	–	36.9[a]	40.83	44.61
Remaining life-time to be expected at the age of 45 (in years)				
Hungary	23.73	25.26	18.31[b]	24.10
Austria	–	24.7[a]	27.31	30.67

a) 1930–1933

b) The longest life spans

Table 2
Male Mortality in the Age-Group 30–69 in Hungary and Austria Between 1920–21 and 1993

Age	1920–1921	1930	1967		1993	
	Hungary	Hungary	Hungary	Austria	Hungary	Austria
	Number of deaths per thousand					
30–34	7.53	5.26	2.01	2.12	3.20	1.21
35–39	8.60	6.45	2.55	2.83	5.95	1.89
40–44	10.01	7.97	3.74	3.88	9.13	2.74
45–49	12.74	10.15	5.41	5.92	12.97	4.49
50–54	16.83	13.57	8.68	9.57	19.58	7.11
55–59	23.61	18.98	14.44	15.96	25.67	11.08
60–64	33.89	27.57	24.68	28.07	36.36	18.58
65–69	52.38	43.21	39.45	45.08	49.43	28.13

In the mid-sixties a downward tendency in mortality lasting for decades was broken. Rising mortality is caused mostly by non-contagious diseases developing insidiously through decades, like cancer and various coronary diseases, and by violent death. In the last two and a half decades, mortality has been rising rapidly in age groups where certain chronic diseases may already cause death, and where violent death is more common. In most cases, symptom-free decades are followed by a shorter period of the illness full of symptoms and complaints, and the pathological process leads inevitably to death. Pulmonic plague can kill a person in twenty-four hours, cigarettes may need twenty-four years to reach the same result, still both plague and cancer may destroy the lungs. It is far from being an exaggeration to say that smoking, alcoholism and the unhealthy diet are the fatal horsemen of the apocalypse in Hungary on the threshold of the third millennium.

It is difficult to assess the damage done by this epidemiological situation to the economy in general, since the benefits for the producers and dealers of tobacco and alcoholic drinks are, at the same time, very significant, and the budget also profits by these products through various taxes. Also, thousands of people depend on the production of tobacco and wine for their livelihood in agriculture, to say nothing of the production and marketing of cigarettes, wine, and spirits.

If growing numbers of people live to an old age, and there is a growing number of disabled on social security, an ever greater amount of money will have to be spent on pension, and the costs of health care will rise. So there are conflicting interests to lowering the mortality rates.

Table 3
Number of Deaths and Changes in the Mortality Rate, 1946–1951 to 1991–1994

Period	Number of deaths per year	Mortality per thousand
1946–1950	114,284	12.5
1951–1955	106,800	11.1
1956–1960	102,230	10.3
1961–1965	102,701	10.2
1966–1970	112,737	11.0
1971–1975	124,457	11.9
1976–1980	137,315	12.9
1981–1985	146,408	13.7
1986–1990	144,017	13.8
1991–1994	147,682	14.3

The rise of the mortality rates is mainly due to the aging of the population, but it is influenced also by the fact that certain age groups, primarily the one between thirty-five and sixty-four, show higher mortality than before.

In 1994, life expectancy of the total Hungarian population was 69.4 years. In the early 1990s, the average life expectancy of people living in the industrial countries was 74.0 years, but in Japan it reached 79.5 and in Austria 76.0. So life expectancy for the Hungarian people lags ten years behind Japan, which is first in rank in the world. The difference between Hungary and Austria in this respect in the early 1990s was seven years to the advantage of the latter. The Hungarian life expectancy of 69.4 years is one of the lowest known figure among the developed countries.

It is important to note that even the slightest decrease in life expectancy is symptomatic, since in the second half of the twentieth century it occurred permanently only at places where famine and war decimated the population. It has not occurred in any industrial country with developed health services, except the former Soviet Union and most of its successor states.

In the 1980s and in the early 1990s life expectancy calculated for the total population consolidated at a low level by European standards, and fluctuated between 69 and 70. It was the lowest in 1983, and the highest in 1988. Recently it has decreased again, and shows a slight rise only in 1994.

Table 4

Life Expectancy at Birth, 1900–1901 to 1994

Year	Life expectancy at birth in years
1900–1901	37.33
1920–1921	42.05
1930–1931	50.21
1941	56.54
1949	61.36
1960	68.03
1966	69.93
1970	69.20
1980	69.02
1981	69.11
1982	69.34
1983	68.95
1984	69.01
1985	69.00
1986	69.19
1987	69.64
1988	70.05
1989	69.53
1990	69.33
1991	69.31
1992	69.00
1993	69.02
1994	69.39

Since male and female mortality differs significantly, life expenctancy at birth is usually calculated separately for both sexes.

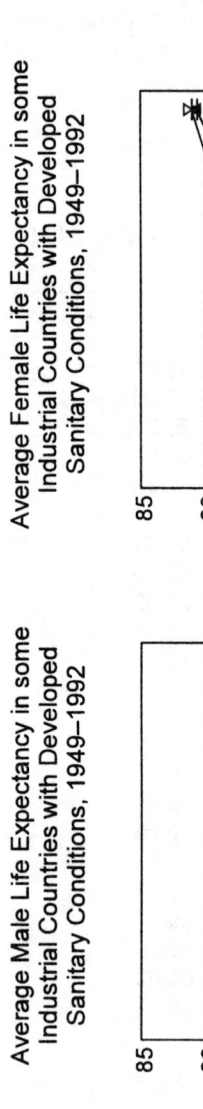
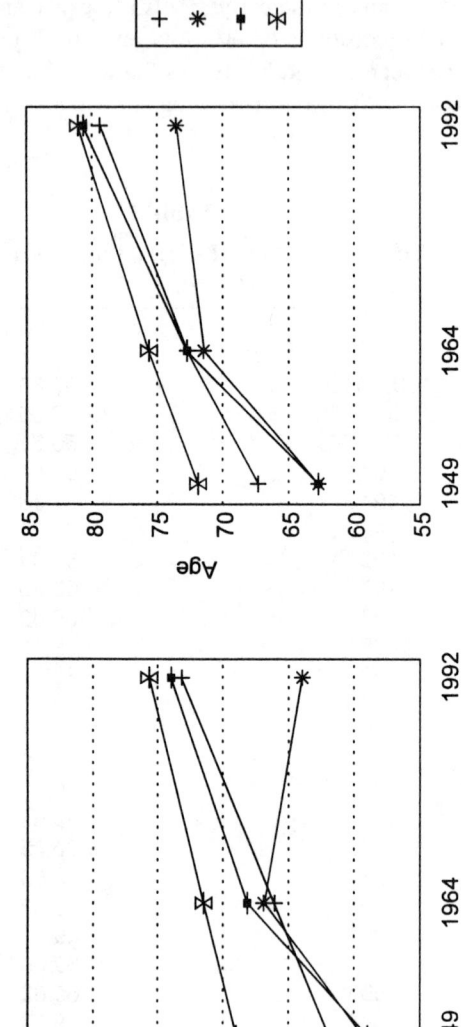

The following chart shows the remaining years to be expected by the 45-year-old male population of certain countries in the first half of the 1990s.

Country	Years
Japan (1992)	33.3
Switzerland (1992)	32.5
Sweden (1990)	32.1
Canada (1991)	32.1
France (1991)	31.9
Spain (1990)	31.7
United States of America (1990)	30.8
Austria (1992)	30.7
Bulgaria (1992)	27.2
Romania (1992)	26.5
Czechoslovakia (1991)	26.2
Poland (1992)	26.1
Ukraine (1990)	25.8
Hungary (1992)	24.3

Life expectancy for middle-aged Hungarian women is also the worst in international comparison:

Country	Years
Japan (1992)	39.3
France (1991)	38.8
Switzerland (1992)	38.3
Canada (1991)	37.6
Spain (1990)	37.3
Sweden (1990)	37.0
United States of America (1990)	36.2
Austria (1992)	36.1
Poland (1992)	33.1
Czechoslovakia (1991)	32.7
Ukraine (1990)	32.6
Bulgaria (1992)	32.5
Romania (1992)	31.8
Hungary (1992)	31.5

Male mortality is higher than female mortality in all industrial countries. In the past decade male mortality was 15.2–16.0 per thousand, while female mortality was only 12.6–12.8 per thousand. The difference is, however, even greater if one takes the age distribution of the male and female population as identical. In that case, the mortality rate of men was 16.8–17.0, while that of women was merely 9.9–9.3 per thousand in the 1980s and in the early 1990s. The standardized mortality rate of men in 1994 was 84 per cent higher than that of women.

Although higher male mortality can be attributed also to biological factors, it is mainly due to other causes. As regards biology, the rudimentary nature of the twenty-third chromosome determining sex in men is probably to blame for the different chances of the two sexes. The difference due to this factor is thought to be two years to the detriment of men.

However, the difference in life expectancy in most developed industrial countries is much greater than that: it can even exceed a decade. In Hungary it was 9.4 years in 1994. This means that at least seven years of difference can be attributed to non-biological causes. This phenomenon is generally explained by the changes in the structure of illnesses and the distribution of the causes of death to the favour of women. The gap between the life expenctancies of the two sexes has widened significantly in the course of the twentieth century. At the turn of the century, men expected to live only 1.6 years less than women, and actual life expectancy was mainly determined by mortality caused by infectious diseases and by death due to childbirth. Today it is determined mostly by the spread of chronic, non-infectious diseases like heart disease, cancer, cerebro-vascular diseases, and by accidents. These causes of death are mostly dependent on the lifestyle of the person involved. Men are more often addicted to harmful habits endangering their health like smoking and excessive drinking than women. Traffic and other accidents, and suicide, are also more common among men. The combination of all these factors contribute to a much higher mortality in men.

Table 5
The Number of Deaths and the Mortality Rate of Both Sexes, and the Ratio of Male and Female Mortality, 1948–1994

Year	Number of deaths		Number of deaths per 1,000 inhabitants		Standardized number of deaths per 1,000 inhabitants[a]		Ratio of male and female mortality in per cent	
	Male	Female	Male	Female	Male	Female	Rough	Standardized
1948	55,797	49,983	12.68	10.50	17.67	13.87	120.76	127.38
1960	51,667	49,858	10.70	9.65	15.20	11.39	110.88	133.48
1970	62,545	57,652	12.53	10.84	15.14	10.60	115.59	142.87
1980	76,729	68,626	14.79	12.43	16.64	10.27	118.99	162.10
1981	76,500	68,257	14.75	12.35	16.45	10.09	119.43	162.99
1982	76,220	68,098	14.71	12.33	16.44	9.95	119.30	165.13
1983	78,651	69,992	15.21	12.68	16.97	10.15	119.95	167.19
1984	78,239	68,470	15.17	12.42	16.80	9.83	122.14	170.85
1985	78,034	69,580	15.17	12.64	16.79	9.89	120.02	169.88
1986	77,059	70,030	15.01	12.74	16.54	9.86	117.82	167.80
1987	74,917	67,684	14.63	12.33	16.07	9.45	118.65	170.02
1988	73,339	66,703	14.35	12.16	15.66	9.21	118.01	170.02
1989	76,521	68,174	15.00	12.45	16.25	9.33	120.48	174.16
1990	76,936	68,724	15.45	12.76	16.71	9.56	121.08	174.86
1991	76,762	68,051	15.46	12.65	16.67	9.42	122.21	176.98
1992	79,633	69,148	16.08	12.87	17.26	9.53	124.94	181.17
1993	80,498	69,746	16.32	13.01	17.30	9.58	125.44	180.58
1994	78,654	68,235	16.01	12.76	17.02	9.25	125.47	184.00

a) Data standardized according to age-groups suggested by the WHO for the European population.

Table 6
Male and Female Life Expectancy at Birth and Their Differences, 1900–1901 – 1994

Year or average of years	Male	Female	Difference of male and female life expectancy at birth
	life expectancy		
	in years		
1900–1901	36.56	38.15	1.59
1920–1921	41.04	43.12	2.08
1941	54.95	58.24	3.29
1949	59.28	63.40	4.12
1960	65.89	70.10	4.21
1966	67.53	72.23	4.70
1970	66.31	72.08	5.77
1980	65.45	72.70	7.25
1981	65.46	72.86	7.40
1982	65.63	73.18	7.55
1983	65.08	72.99	7.91
1984	65.05	73.16	8.11
1985	65.09	73.07	7.98
1986	65.30	73.21	7.91
1987	65.67	73.74	8.07
1988	66.16	74.03	7.87
1989	65.44	73.79	8.35
1990	65.13	73.71	8.58
1991	65.02	73.83	8.81
1992	64.55	73.73	9.18
1993	64.53	73.81	9.28
1994	64.84	74.23	9.39

Annually about fifteen thousand deaths out of the mortality surplus of men in the nearly three decades between 1966 and 1994 resulted from the actual deterioration of mortality, while about twelve thousand deaths could be attributed to the aging of the population. The former meant fifty-six per cent, and the latter forty-four. While the deterioration of certain age-specific mortality rates of the male population was so significant that it led to lower life expenctancy at birth, and an even lower expectancy in adulthood, which practically meant the rise of standardized mortality rates, female mortality remained unchanged. Even if mortality rose sometimes in certain age groups, this did not lead to a decrease in life expectancy or to the rise of standardized mortality rates. On the contrary, the annual mortality surplus of about nineteen thousand deaths in the female population in the last three decades can be attributed fully to old age. What is more, without the improvement of certain age-specific mortality rates, this surplus would be twenty-three thousand.

The Age Structure of Mortality

It is a very favorable phenomenon that people today usually die in old age, and infant mortality is very rare compared to the absolute numbers and rates four decades ago. It gives, however, cause for serious alarm that the relative number of deaths in the age group 15 to 39 grew more in the past decade than would have been justified by the ratio of the age group within the total population. Similarly, the rate of deaths in the age group 40 to 59 is greater now than it was ten years ago, and even more than in 1949, although the rate of this age group within the population rose only slightly between 1980 and 1994.

Table 7
Infant Mortality, 1947–1994

Year	Infant mortality per 1,000 live births				Among those born with		As a result of	
	0–364	0–27	28–364	Perinatal mortality[b]	less than 2,500 grams	2,500 grams or more	congenital causes	acquired causes
		days of life[a]						
1947	106.6	49.1	57.6	50.9	–	–	46.0	60.6
1955	60.0	31.2	28.8	38.7	291.6	39.1	33.0	27.0
1964	40.0	26.7	13.3	33.8	227.8	18.3	28.5	11.5
1970	35.9	28.4	7.5	34.5	229.8	12.7	27.9	8.0
1980	23.2	17.9	5.3	23.1	152.7	8.2	18.9	4.2
1981	20.8	15.7	5.1	21.2	133.7	8.0	16.5	4.3
1982	20.0	15.5	4.5	20.3	133.4	7.5	16.1	3.9
1983	19.0	14.2	4.8	18.3	123.8	7.7	15.0	4.0
1984	20.4	15.8	4.6	18.9	142.0	6.8	16.3	4.1
1985	20.4	15.7	4.7	19.0	141.3	7.0	16.5	3.8
1986	19.0	14.7	4.3	18.3	130.4	6.9	16.2	2.9
1987	17.3	13.0	4.3	17.4	122.8	6.1	14.5	2.8
1988	15.8	11.8	4.0	15.9	112.0	5.9	13.2	2.6
1989	15.7	11.7	4.0	14.6	114.0	5.8	13.2	2.5
1990	14.8	10.8	4.0	14.3	106.1	5.5	12.1	2.7
1991	15.6	11.4	4.3	13.5	114.5	5.5	13.5	2.6
1992	14.1	10.2	3.9	12.1	106.0	5.0	11.9	2.2
1993	12.5	8.6	3.9	10.2	90.2	5.1	10.0	2.5
1994	11.5	7.9	3.6	9.3	84.3	4.7	9.5	2.1

a) Between 1947 and 1957 deaths occurring in the first month after birth were called neonatal, and the ones occurring in the rest of the first year postneonatal.
b) Those dying after the twenty-eighth completed week of pregnancy or in the first six days after birth.

Changes in Mortality in Hungary between 1980 and 1994   123

Table 8
Rate of Male Mortality According to Age-Groups, 1920-1921 – 1994

Age group	Year (average of years)											
	1920-1921	1930-1931	1938	1948-1949	1959-1960	1969-1970	1979-1980	1989	1990	1991	1992	1993	1994

Number of deaths per thousand in an age group

Age group	1920-1921	1930-1931	1938	1948-1949	1959-1960	1969-1970	1979-1980	1989	1990	1991	1992	1993	1994
0	207.19	170.74	144.04	101.97	55.13	39.29	26.43	17.08	16.43	17.34	15.48	13.72	12.79
1	50.66	31.26	23.78	9.11	4.13	1.87	1.13	0.91	0.74	1.09	0.77	0.81	0.78
2	38.15	10.85	7.77	4.28	1.59	0.97	0.69	0.58	0.51	0.45	0.67	0.57	0.61
3-4	16.24	6.35	4.22	2.54	0.83	0.66	0.45	0.48	0.49	0.41	0.29	0.42	0.34
0-4	103.10	52.80	41.58	28.38	11.42	9.41	5.24	3.89	3.80	4.06	3.49	3.09	2.88
5-9	7.34	3.43	2.37	1.51	0.58	0.46	0.41	0.32	0.31	0.35	0.26	0.27	0.20
10-14	4.18	2.38	1.90	1.20	0.62	0.47	0.43	0.33	0.33	0.28	0.27	0.27	0.30
15-19	6.41	4.20	3.38	2.22	1.20	1.06	1.05	0.98	1.01	0.84	0.92	0.70	0.79
20-24	8.58	5.81	5.40	3.68	1.60	1.45	1.47	1.40	1.55	1.42	1.49	1.28	1.23
25-29	7.57	5.43	4.56	3.72	1.62	1.68	1.73	1.85	2.03	2.10	1.88	1.81	1.57
30-34	7.52	5.55	4.62	4.03	1.99	2.04	2.29	3.03	3.23	3.06	3.39	3.20	3.10
35-39	8.58	6.78	5.41	4.61	2.54	2.72	3.71	4.58	4.90	4.90	5.46	5.95	5.83
40-44	10.00	8.35	7.46	6.26	3.48	4.23	5.87	6.82	7.10	7.77	8.63	9.13	9.08
45-49	12.72	10.48	10.00	8.40	5.50	6.12	8.74	11.05	11.51	11.70	12.73	12.97	13.12
50-54	16.82	14.28	14.00	11.83	9.14	9.53	13.65	16.60	16.74	17.39	18.76	19.58	18.82
55-59	23.59	19.92	19.62	17.27	16.08	14.78	20.10	23.27	24.59	24.14	25.77	25.67	26.19
60-64	33.85	29.24	29.36	25.32	25.30	25.80	29.65	34.10	34.15	34.07	34.87	36.36	34.91
65-69	52.32	45.23	46.13	37.52	40.02	42.68	43.48	46.87	46.28	46.61	50.1	49.43	48.22
70-74	84.72	71.83	70.92	60.86	63.35	66.49	69.87	64.10	63.28	63.04	65.8	67.69	68.13
75-79	136.56	114.62	114.95	97.18	101.79	104.51	109.59	97.42	100.45	99.94	98.49	102.32	93.35
80-84	216.09	196.58	182.56	154.69	162.29	159.00	161.76	147.69	152.50	148.10	15.87	147.78	144.71
85-X	278.58	292.36	271.94	257.39	271.33	269.11	260.10	232.88	252.95	247.54	246.92	249.20	239.69
Total	22.50	16.94	15.05	12.47	10.93	12.33	14.34	15.00	15.45	15.46	16.08	16.32	16.01

Table 9
Rate of Female Mortality According to Age-Groups, 1920–1921 – 1994

Age group	1920–1921	1930–1931	1938	1948–1949	1959–1960	1969–1970	1979–1980	1989	1990	1991	1992	1993	1994
							Number of deaths per thousand in an age-group						
0	177.31	142.38	117.88	82.44	44.65	32.07	20.58	14.33	13.15	13.85	12.60	11.12	10.23
1	47.73	29.71	22.27	8.03	3.69	1.76	0.97	0.77	0.79	0.63	0.67	0.69	0.85
2	38.66	10.53	6.59	4.03	1.45	0.78	0.52	0.46	0.40	0.46	0.52	0.46	0.38
3–4	15.54	6.25	3.85	2.39	0.69	0.54	0.31	0.35	0.28	0.29	0.27	0.37	0.22
0–4	87.72	43.82	33.94	22.72	9.22	7.69	4.11	3.22	3.03	3.20	2.82	2.51	2.29
5–9	7.27	3.62	2.16	1.33	0.40	0.31	0.24	0.24	0.27	0.21	0.20	0.17	0.17
10–14	4.83	2.61	1.97	1.09	0.33	0.28	0.24	0.21	0.21	0.19	0.22	0.23	0.15
15–19	6.92	4.81	3.58	2.04	0.59	0.41	0.43	0.34	0.44	0.39	0.37	0.35	0.31
20–24	8.13	6.05	4.72	2.94	0.79	0.53	0.51	0.51	0.57	0.53	0.47	0.41	0.46
25–29	8.24	5.94	4.33	2.86	0.99	0.67	0.68	0.73	0.66	0.64	0.65	0.63	0.52
30–34	8.17	5.64	4.23	2.97	1.28	0.95	1.04	1.24	1.26	1.22	1.30	1.26	1.09
35–39	8.44	6.25	4.80	3.29	1.84	1.45	1.62	1.87	1.94	2.07	2.15	2.40	2.20
40–44	8.95	6.83	5.83	4.01	2.72	2.27	2.74	2.89	3.03	2.88	3.61	3.43	3.26
45–49	10.80	8.42	7.66	5.60	3.85	3.59	4.21	4.37	4.45	4.59	4.61	4.84	4.67
50–54	13.90	11.67	10.38	7.83	5.93	5.82	6.33	6.50	6.51	6.50	6.81	6.79	6.77
55–59	20.04	16.42	15.58	11.63	9.41	8.43	9.63	9.91	9.61	9.56	9.97	10.09	9.56
60–64	30.33	24.76	23.72	18.17	15.82	13.95	15.63	14.42	14.62	14.66	14.50	15.14	14.73
65–69	48.92	40.88	38.63	29.81	28.07	24.90	23.86	23.16	23.24	23.05	23.10	23.37	22.32
70–74	78.87	64.62	64.57	50.44	48.98	44.78	41.98	38.43	37.37	35.96	36.34	36.51	37.96
75–79	135.49	108.73	107.31	85.30	86.92	77.91	72.89	63.75	66.56	64.68	67.32	65.97	62.01
80–84	200.61	168.15	166.63	136.19	145.96	131.31	123.30	111.46	113.93	110.99	111.81	110.19	106.77
85–X	254.24	253.09	271.24	226.39	239.91	227.18	219.16	196.62	212.95	210.94	210.21	211.79	203.49
Total	20.10	15.24	13.48	10.58	9.73	10.67	12.08	12.45	12.76	12.65	12.87	13.01	12.76

The Causal Structure of Mortality

Of the several thousand catalogued diseases, only ten to fifteen cause the majority of deaths.

Most people die of circulatory diseases, with tumours as second and violent death as third in the list of causes. These three groups make up seventy-five to eighty-five per cent of the total number of deaths. The diseases of the digestive system are the fourth in the row, and those of the respiratory system are the fifth. The latter two main groups may interchange.

The structure of the causes of mortality in Hungary and in the countries with the lowest mortality rates do not differ significantly. The only difference lies in the frequency of the various causes and in the rate of premature deaths within some groups of causes.

Mortality due to infectious diseases is steadily and significantly decreasing and these diseases became insignificant from an epidemiological point of view by the early 1990s. They make up a mere 0.6 per cent of the total deaths. Their actual rate is, however, larger than this, because influenza and pneumonia do not belong to this group. Should they belong there, the infectious diseases would make up 1.4 per cent of the total number of deaths. In the years with serious influenza epidemics, this rate could be around four per cent.

The number of deaths due to most infectious diseases will most probably decrease in the future, but AIDS will kill more and more people each year. The spread of AIDS and the appearance of bacteria resistant to medicaments against tuberculosis will contribute to a growing mortality owing to tuberculosis. The growing number of people living in extreme poverty will also contribute to the spread of tuberculosis. So the next decade will most probably witness a rise in the number of deaths due to infectious diseases compared to the 1980s and the early 1990s.

The relative significance of neoplasms has also risen in the last decade. The increasing number of deaths owing to lung cancer, cancers of the oral cavity, the pharynx, the oesophagus, the colon, the rectum, and the female breast will contribute to the greater share of this main group among causes of death.

Mortality due to the circulatory system is probably smaller than the data would suggest. When diagnosing old people, doctors tend to attribute their problems to arteriosclerosis, so the aging of the population may increase the share of this group, though the hopefully decreasing death rate of hypertension and the cerebro-vascular diseases, as well as the more correct diagnoses may result in a lower rate for this group. The number of death due to heart disease cannot, however, be assessed, which makes the prediction of mortality in this group rather difficult. It is still probable that it will lose its relative weight in mortality rates in the future.

No definite trend can be established in connection with mortality due to respiratory diseases. The relative significance of this group is due mainly to influenza, pneumonia and obstructive lung diseases like chronic bronchitis, emphysema, and asthma. (The inclusion of influenza and pneumonia in this group is, however, questionable.)

The recurring epidemics of influenza are sure to raise mortality every third or fifth year. The number of deaths due to pheumonia mostly depends on these epidemics. Diagnosing obstructive lung diseases as causes of death can be very subjective. So in this group no noticeable trend is to be expected in the future.

Mortality due to illnesses of the digestive system has been rising in the last few decades, and this is to be expected for the future, too. The case is similar for alcoholic cirrhosis, but alcoholism will raise the number of deaths due to illnesses of the pancreas as well. The after-effects of virus hepatitis B are also likely to contribute to the rise of mortality rates.

Although the rate of violent death has been steadily growing for decades, it is this group where the possibilities of decreasing mortality are the greatest. Realizing these possibilities also seems relatively easy. The number of violent deaths will hopefully decrease in the near future. Its relative significance within the total number of deaths is determined by three main categories, namely, by accidents with motor vehicles, suicide, and falls. Seventy-three per cent of the latter happens to people of sixty years and over, fifty-two per cent to people of eighty and over. Most of these people do not fall from a higher place but stumble and fall on an even floor. Eighty per cent of these falls happen at home, the patients suffering fracture of the femoral neck, and dying of the complications. This is not the kind of violent death as, for example, a car accident. This type of accident is more related to the physical and mental state of the elderly people, and it would be more reasonable to define it as a cause of death related to old age.

Mortality due to the rest of the causes has steadily been decreasing and is sure to decrease in the future, too.

Table 10
Distribution of Mortality According to the Main Divisions Causing Death, 1947–1994

Total population

Main divisions of the causes of death	1947a	1964a	1980a	1990a	1991a	1991b	1992a	1992b	1993a	1993b	1994a	1994b
I. Contagious diseases and diseases caused by parasites	11.8	3.3	1.1	0.7	0.7	1.3	0.6	1.4	0.6	1.5	0.6	1.4
II. Tumours	9.3	19.0	19.2	21.4	21.9	21.9	22.0	22.0	21.7	21.7	22.5	22.5
VII. Illnesses of the circulatory system	23.1	50.9	52.9	52.4	52.4	45.6	51.3	43.7	51.5	45.2	50.5	44.2
VIII. Illnesses of the respiratory sytem	11.5	4.3	6.9	4.6	4.3	3.7	4.7	3.9	4.7	3.8	4.7	3.9
IX. Illnesses of the digestive system	6.2	3.5	4.8	6.2	6.3	6.3	7.3	7.3	8.0	8.0	8.2	8.2
XVI. Symptoms and insufficiently diagnosed statuses	–	–	–	–	–	8.7	–	9.6	0.1	8.3	0.1	8.2
E. Outward causes of injury and poisoning	5.3	6.9	8.5	9.1	8.9	7.0	8.8	6.7	8.2	6.3	8.1	6.3
Others	32.8	12.1	6.6	5.6	5.5	5.5	5.4	5.4	5.2	5.2	5.3	5.3
Total	100.0	100.0	100.0	100.0	100.0	100.0	100.0	100.0	100.0	100.0	100.0	100.0

a) According to the ninth revision of BNO in 1975.
b) Following a certain regrouping of the illnesses included in the ninth revision of BNO.

Table 10
Distribution of Mortality According to the Main Divisions Causing Death, 1947–1994
(continued)

Main divisions of the causes of death	1947[a]	1964[a]	1980[a]	1990[a]	1991[a]	1991[b]	1992[a]	1992[b]	1993[a]	1993[b]	1994[a]	1994[b]
	Male population											
I. Contagious diseases and diseases caused by parasites	12.6	4.3	1.3	0.8	0.8	1.4	0.8	1.6	0.8	1.7	0.8	1.6
II. Tumours	8.1	18.9	20.0	22.9	23.5	23.5	23.2	23.2	22.6	22.6	23.7	23.7
VII. Illnesses of the circulatory system	21.7	46.4	48.9	47.4	47.2	42.2	46.0	40.1	46.3	41.9	45.0	40.6
VIII. Illnesses of the respiratory sytem	12.2	5.0	8.0	5.3	5.1	5.1	5.3	4.5	5.4	4.6	5.3	4.5
IX. Illnesses of the digestive system	6.7	3.8	5.3	7.2	7.5	7.5	8.9	8.9	9.7	9.7	10.2	10.2
XVI. Symptoms and insufficiently diagnosed statuses	–	–	–	–	–	6.1	–	7.4	0.1	5.7	0.1	5.6
E. Outward causes of injury and poisoning	7.4	9.1	10.2	11.0	10.7	9.6	10.6	9.0	9.9	8.7	9.8	8.7
Others	31.2	12.6	6.3	5.4	5.2	4.6	5.2	5.2	5.1	5.1	5.2	5.2
Total	100.0	100.0	100.0	100.0	100.0	100.0	100.0	100.0	100.0	100.0	100.0	100.0

a) According to the ninth revision of BNO in 1975.
b) Following a certain regrouping of the illnesses included in the ninth revision of BNO.

Table 10
Distribution of Mortality According to the Main Divisions Causing Death, 1947–1994
(continued)

Female population

Main divisions of the causes of death	1947a	1964a	1980a	1990a	1991a	1991b	1992a	1992b	1993a	1993b	1994a	1994b
I. Contagious diseases and diseases caused by parasites	10.9	2.2	0.8	0.5	0.5	1.0	0.5	1.3	0.4	1.3	0.4	1.2
II. Tumours	10.8	19.1	18.3	19.8	20.1	20.1	20.6	20.6	20.5	20.5	21.1	21.1
VII. Illnesses of the circulatory system	24.6	55.6	57.5	58.1	58.1	49.5	57.3	47.7	57.5	48.9	56.9	48.4
VIII. Illnesses of the respiratory sytem	10.7	3.6	5.6	3.7	3.5	3.5	3.9	3.1	3.9	3.0	4.0	3.2
IX. Illnesses of the digestive system	5.7	3.3	4.3	5.1	5.1	5.1	5.5	5.5	5.9	5.9	6.0	6.0
XVI. Symptoms and insufficiently diagnosed statuses	–	–	–	–	–	11.6	–	12.3	0.1	11.4	0.1	11.2
E. Outward causes of injury and poisoning	2.9	4.7	6.5	7.0	6.9	4.0	6.7	4.0	6.2	3.6	6.1	3.4
Others	34.5	11.6	7.0	5.8	5.8	5.2	5.5	5.5	5.4	5.4	5.4	5.4
Total	100.0	100.0	100.0	100.0	100.0	100.0	100.0	100.0	100.0	100.0	100.0	100.0

a) According to the ninth revision of BNO in 1975.
b) Following a certain regrouping of the illnesses included in the ninth revision of BNO.

Table 11

Mortality Caused by the Most Frequent Illnesses Leading to Death, 1947–1994

Main divisions of the causes of death	Standardized mortality rate according to the age distribution of the European population as suggested by WHO							
	1947	1964	1980	1990	1991	1992	1993	1994
	Mortality per 10.000 inhabitants							
I. Contagious diseases and diseases caused by parasites	14.99	3.47	1.34	0.86	0.84	0.85	0.74	0.79
II. Tumours	15.80	20.39	24.34	27.00	27.41	28.21	28.12	28.29
VII. Illnesses of the circulatory system	41.99	60.34	68.89	64.47	63.80	63.97	64.74	61.38
VIII. Illnesses of the respiratory sytem	16.66	4.97	8.95	5.71	5.38	5.93	6.05	5.80
IX. Illnesses of the digestive system	8.13	3.85	6.30	8.15	8.34	9.90	10.94	11.08
E. Outward causes of injury and poisoning	7.51	7.47	11.47	12.12	11.75	11.89	11.18	10.75
Others	62.16	14.29	9.05	7.79	7.64	7.68	7.81	7.62
Total	167.24	114.77	130.34	126.11	125.17	128.43	129.59	125.71

a) According to the ninth revision of BNO in 1975.

Table 11
Mortality Caused by the Most Frequent Illnesses Leading to Death, 1947–1994
(continued)

Main divisions of the causes of death	Standardized mortality rate according to the age distribution of the European population as suggested by WHO								
	1947	1964	1980	1990	1991	1992	1993	1994	
			Mortality per 10.000 male inhabitants						
I. Contagious diseases and diseases caused by parasites	18.10	5.17	2.09	1.38	1.34	1.36	1.41	1.30
II. Tumours	16.22	23.78	31.70	37.32	38.10	39.24	38.83	39.36
VII. Illnesses of the circulatory system	45.92	66.48	84.34	80.74	80.49	81.02	82.58	78.07
VIII. Illnesses of the respiratory system	19.91	6.54	13.70	9.14	8.72	9.43	9.68	9.21
IX. Illnesses of the digestive system	9.52	4.72	8.56	11.55	12.03	14.73	16.35	16.70
E. Outward causes of injury and poisoning	11.90	10.59	16.24	17.71	17.14	17.59	16.61	16.18
Others	66.63	16.69	10.27	9.21	8.83	9.25	7.55	9.37
Total	188.18	133.97	166.90	167.05	166.65	172.62	173.01	170.20

a) According to the ninth revision of BNO in 1975.

Table 11
Mortality Caused by the Most Frequent Illnesses Leading to Death, 1947–1994
(continued)

Main divisions of the causes of death	Standardized mortality rate according to the age distribution of the European population as suggested by WHO							
	1947	1964	1980	1990	1991	1992	1993	1994
	Mortality per 10.000 female inhabitants							
I. Contagious diseases and diseases caused by parasites	12.33	2.09	0.79	0.47	0.46	0.45	0.22	0.38
II. Tumours	15.52	17.88	19.16	19.75	19.89	20.47	20.59	20.55
VII. Illnesses of the circulatory system	38.77	55.55	57.42	52.51	51.64	51.45	51.77	49.34
VIII. Illnesses of the respiratory sytem	13.93	3.73	5.82	3.58	3.29	3.75	3.72	3.65
IX. Illnesses of the digestive system	6.89	3.14	4.50	5.38	5.33	5.88	6.48	6.36
E. Outward causes of injury and poisoning	3.70	4.62	7.17	7.15	6.91	6.88	6.41	6.02
Others	58.40	12.35	8.15	6.66	6.62	6.40	6.56	6.18
Total	149.54	99.37	103.01	95.51	94.14	95.28	95.75	92.47

a) According to the ninth revision of BNO in 1975.

Causes of Middle-Aged Male Mortality in the Age Group 50–54

The worsening of mortality rates in general in the past decades can be seen in the changing mortality rates of men between 50 and 54 in the given period, especially if one contrasts it to similar data in a neighboring country.

Table 12

Number of Deaths per 100,000 Men in the Age-Group 50–54

Year	Hungary	Austria
1920–1921	1682	–
1960	886	1058
1980	1422	962
1990	1674	674
1993	1958	711

Male mortality per 100,000 in this age group was more than double between 1960 and 1994, 83.4 per cent of which could be attributed to increase in only six categories. These are the following:

Taking the total growth between 1960 and 1994 as 100 per cent: 30.6 per cent of the total growth was due to cirrhosis of the liver, 14.4 per cent to coronary diseases, 14.2 per cent to lung cancer, 8.4 per cent to cancers of the lips, the oral cavity, the pharynx, and the oesophagus, 8.1 per cent to cerebral diseases, and 4.0 per cent to suicide.

Between 1960 and 1994 the increase of deaths due to cirrhosis contributed dramatically to the growth of mortality in the male age group of 50 to 54. There is no other illness that would even approximate the rapid increase of mortality than this.

In the decade and a half between 1980 and 1994 male mortality in the age group 50–54 rose by thirty-two per cent. To judge the contribution of the most frequent causes of death to this increase, one can take the total increase for 100 per cent and examine the relative weight of the individual group accordingly. 82.7 percent of the rise in mortality rates goes back to only three illnesses: 52.6 per cent of the total increase between 1980 and 1994 was due to cirrhosis of the liver, 17.2 per cent to lung cancer, and 12.9 per cent to cancers of the lips, the oral cavity, the pharynx and the oesophagus.

Mortality due to coronary diseases remained practically unchanged in the given period, while mortality due to suicide decreased by 10 per cent.

The risks contributing to these illnesses are well known. Alcoholism and smoking are most probaby to blame for the greatest number of deaths in this group. Conservative estimates maintain that 17 per cent of the total number of deaths can be attributed to the former factor, and 16 per cent to the latter. However, it seems probable that every fifth death in the male age group of 50–54 can be related to alcoholism, and mortality due to smoking can also approximate 18 per cent. So 33–38 per cent of male mortality in the age group 50–54 is thought to be the result of excessive drinking and smoking. In the case of women in the same age group, mortality attributed to these two causes amount "only" to 19–20 per cent.

Recently, mortality due to alcoholism has been outpacing mortality from smoking in the above mentioned age group. At the same time, it would be very unsafe to estimate the effect of an unhealthy diet on the increase of mortality.

The analysis of the causes of death has furnished another lesson, namely that about one fifth of the deaths could have been prevented by proper medical intervention. Mostly illnesses due to hypertension and cerebro-vascular diseases belong to this group.

The available data show that nowhere in the world, not even in the countries of the "third world" is the mortality of middle-aged men as high as in Hungary. Mortality in rural areas is even more unfavourable than the national average: male mortality in the age group of 50–54 exceeds the national average by 16 per cent. The socially determined differences are still greater than the geographical ones.

Conclusions

> *"Epidemics are actually the consequences of social problems."*
> Virchow

It is difficult to determine the relative responsibility of the individuals and the society. An analysis concentrating on the possibilities of improving the state of health of the population deals primarily with the responsibilities of society, and with the causes of the rapid increase of mortality going back to a definite group of medical entities when the successful western societies show just the opposite trend.

There is no satisfactory explanation for the deterioration of the state of health of the Hungarian population, and the rise of mortality. No correlation has been established between the two groups of the variables determining this process. One of these groups contains the biological variables, i.e., the direct causes of death for the individual, such as lung cancer, myocardial infarction, stroke, cirrhosis, suicide, etc. The second group is made up of social determinants like poverty, technological backwardness, the lack of organization, low efficiency, etc., that determine the mortality of the whole population. But how is it that the weak cohesion of society eventually leads to the increase of mortality? There must be a third group between the above-mentioned two, linking them and building a bridge between the biological and the social factors. This group is constituted by intermediary variables including risk factors like unhealthy diet damaging health and closely related to the way of life of the individual, smoking, alcoholism, the lack of exercise, excess weight, the damages done to the physical environment like pollution, radioactivity, intensive UV radiation, chemicals used in agriculture and industry, etc. Last but not least, the level of health care also belongs to this group of variables.

Experience shows that the risk factors related to the lifestyle of people play the greatest part in determining the quality of life and the level of mortality in a country.

The social, the intermediary, and the biological factors form a hierarchy, in which the social factors are first in rank. The link between the intermediary and the biological spheres is not dependent on the social regime. Alcoholism leads to cirrhosis no matter what the social and economic environment is like. Such a link can be observed only between the social and the intermediary variables. Therefore, it has to be examined what factors play the greatest part in determining the health and mortality of the Hungarian population, where they differ from the determinants pointed out in a western society, e.g., in Austria, and how they influence the intermediary variables.

In the last three or four decades the cohesion of the family has been shaken as is proved by the growing number of divorces. The disintegrating families leave a great number of mostly young people with disturbed values or without a scale of values altogether undefended against the harmful effects of certain social circumstances.

Rapid vertical mobility played an important part in making alcoholism a widespread disease. Forced industrialization needed labor force, and the unskilled workers coming from the countryside and living in workers' hostels were not only torn out of their original environment, but also left their traditional values behind. Often their relationship with their families was broken, and they were kept at a distance from the Church, too. The trade unions of the various branches of industry urged only for the increase of production and did not care for these workers' interests and human dignity. The big cities did not integrate them and nobody cared for them. Young men in their teens with no order of values or spiritual reserves were influenced by their already destroyed older colleagues. It was almost compulsory to drink, and a drunken state later became a refuge for them.

Ironically enough, the society of the party-state emphasizing the leading role of the working class created a rootless, unprotected and alienated layer of unskilled workers whose unfair chances contributed disproportionately to the high mortality of the country.

Alcoholism can be detected also behind a large number of suicides, traffic and other accidents, the disintegration of the personality, and crimes. Alcoholism is the most significant risk factor underlying most the social maladaptation syndrome.

The twentieth-century history of smoking was largely determined by the two world wars. With the dog-tag, the soldiers were also given their daily ration of cigarettes, which can easily be proved by cohort mortality surveys showing the age-groups that got increasingly contaminated. Female smoking is a relatively new phenomenon. Its effects on mortality can already be seen, but its actual influence on the increase of mortality rates will manifest only later. Just like alcoholism, smoking is also independent of political regimes.

The unhealthy diet of the Hungarian population plays an exceptionally great part in the high mortality level. Traditional Hungarian dishes contain a large amount of cholesterol and much more salt than necessary. Hungarian people eat little fibrous food, and foodstuffs rich in vitamins and containing anti-oxidants are available regularly only for the well-to-do. Besides, too many calories are consumed, and consequent excess weight causes frequent public health problems.

The role of environmental problems in the deterioration of sanitary conditions cannot be assessed for the time being. Pollution must be a major health hazard. The structure of mortality and the great difference between male and female mortality indicates, however, that the effects of environmental pollution are fairly limited. If it were not the differences in the ways of life that determine the changes of mortality but environmental problems, men and women living in the same neighborhood would not differ so much in their mortality.

The health care system is part of a country's infrastructure. Paradoxically, the deterioration of health and the rise of mortality occurred just at the time when every citizen was entitled to get "free of charge and high-quality health care" first as and insurant, later as a personal right. Neither preventive, nor therapeutic healing is effective enough. Many people die a premature death due to illnesses they should not die of, owing to the low efficiency and even the failure of the health care system.

About 18 per cent of all deaths (nearly 28,000 each year) is due to the unsatisfactory health service and could be avoided by careful medical intervention. The similar ratio in the countries of the European Community is 11 per cent, and 10 per cent in the United States of America.

New pieces of knowledge spread very slowly, and this applies also to health promotion. The breakthrough research results in epidemiology pioneered by British and American scholars reach the Hungarian public only slowly, and are known nearly exclusively by professionals and the elite. Without disseminating knowledge on the prevention of illnesses and health promotion on a large scale it cannot be expected that people should take care of themselves according to the demands of the age. Besides illuminating the minds, however, the economic conditions to ensure a healthy way of life should also be created.

Growing unemployment, the impoverishment of the losers of the recent transition, social polarization in general and the changes in the health care system will most probably contribute to further deterioration of the probability of surviving, and to inequality in health in Hungary. If the improvement of public and personal health of the people does not become one of the priorities in this country, the epidemiological crisis may become even more serious.

Pál Péter Tóth

International Migration and Hungary

One thousand and one hundred years ago, prior to their arrival in the Carpathian Basin the Hungarians were typical migrant people. In the thirteenth century much of the population was forced to flee from the Mongol invasion, then in the sixteenth and seventeenth centuries came similar incursions of the Turks. Even the rule of the Habsburgs gave rise to smaller but not less devastating migrations from the first third of the sixteenth century. Finally, there were the two world wars putting thousands on the move. Parallelly, groups of aliens immigrated or were settled down in Hungary in connection with the adoption of Christianity, the establishment of royal power, the specialization of intellectual activities, and various cataclysms of history.

After World War II Hungary became part of the Soviet sphere of interest that came to an end only as late as 1990. The present article deals with the historical antecendents and with the new processes of migration from 1988 linked to international migration in our age. Since that date the following factors contributing to migration may be traced:

- Hungary ceased to be isolated owing to the change of political regimes,
- Hungarians' travelling abroad was no longer restricted by political exigencies,
- Hungary became the destination or a transit station for international migration, partly owing to its geographic position,
- owing to unsolved nationality problems in the neighboring countries and the war in former Yugoslavia 127,208 refugees arrived between 1988 and 1994.

Difficulties in writing this analysis originated from the inaccuracy of the database, the scarcity of experience in this field in Hungary, and from the fact that, owing to the Act on Personal Rights, only part of the data have been made available for me.[1]

Historical Background

In order to be able to understand current processes we have to comprehend changes in citizenship following World War I in this region.

The Dissolution of the Austro-Hungarian Monarchy and the Fate of the Hungarian Citizens

In October 1918 the dissolution of the Austro-Hungarian Monarchy in the wake of the various national aspirations was an accomplished fact. It was, however, not clear at that time how many foreign citizens, Hungarians included, would be situated in the artificially created new nation-states or federations of states.[2]

The peace treaty closing World War I divided the multinational Austro-Hungarian Monarchy into two relatively homogeneous countries, i.e., Austria and Hungary, and into three multinational ones, i.e., Czechoslovakia, the Serbian-Croatian-Slovene Kingdom, and Romania. These countries received much more of the former Hungarian Kingdom both in territory and population than they could hope for on the basis of the principles formulated by the victorious powers.[3]

Changes in the Territory and Population of the Hungarian Kingdom in Consequence of the Peace Treaty Closing World War I[4]

Annexed to	Territory (in square kms)	Population (persons)	Hungarians (persons)
Czechoslovakia	63,004	3,567,575	1,072,000
Romania	102,181	5,236,305	1,664,000
Serbian-Croatian-Slovene Kingdom	21,031	1,519,013	459,000
Austria	4,026	292,588	26,000
Total	190,242	10,615,481	3,221,000

Hungarians who chose to remain Hungarian citizens on disannexed territories as optants or in any other legal status had to leave their homeland, and those remaining there automatically became citizens of Czechoslovakia, Romania, Austria or the Serbian-Croatian-Slovene Kingdom.

As part of their anti-nationality policy, these new states restricted the participation of the Hungarian nationality in the government, their economic activity, and the use of the mother tongue. Even citizenship for many of those who remained was refused, although it was guaranteed by the Peace Treaty. Consequently, hundreds of thousands were forced to flee.[5] The exact number of those forced to migrate in this way is not known. The dimensions of this migration are, however, illustrated by the fact that about 200,000 people came to Hungarian territory from Romania alone in the years 1919 to 1923.

The first modification of the Trianon frontiers prior to World War II took place on November 2, 1938, on the basis of the First Vienna Award. Czechoslovakian territories with predominantly Hungarian population and 90 per cent of the Hungarian nationality living there were reannexed to Hungary. This measure was followed by the reannexation of Sub-Carpathia (Ruthenia) in May, 1939. Northern Transylvania was reannexed in September, 1940, and the Voivodina in the spring of 1941. By 1941, Hungary's territorial increase amounted to 78,680 square kilometers, with an increase of population numbering 5,363,331 of whom hardly more than 50 per cent were Hungarians. It was a special type of international migration with millions taking part without even leaving their homes. The new lines of frontiers involved the resettlement of many who had migrated to Hungary from the disannexed territories, the emigration of those non-Hungarians who had settled down on the annexed territories, and a fresh flight of the nationalities from the divided territories. Following the Second Vienna Award, August 30, 1940, for example, Romanians fled from Northern Transylvania, and Hungarians from Southern Transylvania. This migration involved about 200,000 people respectively.[6]

Ethnic Distribution of the Population of Reannexed Territories, 1938–1941[7]

	Upper Hungary	Sub-Carpathia	Northern Transylvania	Southern Hungary
Hungarians	751,951	63,025	1,343,695	348,840
Germans	17,354	57,435	47,508	178,221
Slovaks	84,905	20,449	20,885	30,153
Romanians	360	11,385	1,069,211	38
Croats	223	141	83	87,994
Serbs	26,171	13	108	150,336
Ruthenes	8,941	342,029	20,622	10,754
Others	66,537	2,010	21,647	136,196
Total	956,442	496,487	2,523,759	942,879

Besides the above described processes, Jewish refugees from Austria, Poland, Slovakia, and Serbia also arrived in Hungary during the period 1938–44, and Hungary gave shelter also to French soldiers and tens of thousands of Polish citizens fleeing from the Germans.[8]

Following the German occupation of Hungary on March 19, 1944 the Germans deported portions of the Gipsy and Jewish population of the country.[9]

The peace treaty closing World War II, however, left the ethnic principle out of consideration again, and besides restoring the situation prior to the Vienna Awards, it gave Czechoslovakia further Hungarian territories.[10] Sub-Carpathia was annexed to the Soviet Union.[11]

Toward the end of World War II, an unprecedented number of people emigrated from Hungary to the West. Some of them left the country in accordance with a government decree from October 20, 1944. Many institutions with all their employees and their family members were transported to the West as the Soviet army was approaching the country. Others left the country out of fear of the Soviet occupation and the possible social and political changes, or of being called to account because of participation in the war. It was also a serious problem that the Hungarian minorities living in the neighboring countries were considered as enemies there, and had to face retribution, physical annihilation, deprivation of rights, forced

International Migration and Hungary 143

labor, or forced resettlement after the war.[12] As a result, nearly 130,000 Hungarians emigrated to Hungary from Romania, about 115–120,000 from Czechoslovakia and Yugoslavia, and more than 70,000 from the Soviet Union.[13]

Statistical data do not make it possible to establish the exact number of those leaving Hungary for the West. Estimates vary between 200,000 and 800,000.[14] Further changes were brought about by the removal of most German ethnic minorities from Hungary (about 250,000 people) after 1945, and by the population exchange between Czechoslovakia and Hungary that meant the loss of about 90,000 people of Slovak nationality.[15]

After the Establishment of Bolshevik-Type Rule in Hungary

Just as in other countries of the Danubian Basin, a Bolshevik-type takeover occurred in Hungary in 1948, giving rise to another wave of migration to the West. Thereafter freedom of movement of the population became limited, and only those could travel abroad who were deemed reliable. The natural flow of international migration was blocked, and illegal emigration was prosecuted.

The Hungarian emigrants leaving the country in 1945 and in 1946–49 were followed in 1956 again by nearly 200,000 people[16] who were joined yearly by 8–10,000 emigrants leaving the country in either legal or illegal ways.[17]

The changes did not leave the number of Hungarians living in various countries of the Danubian Basin unaltered. Besides developments similar to those in Hungary, the anti-minority policy of some neighboring countries played a great part in this process.

Changes in the Number of Hungarians in the Danubian Countries, 1920–91[18]

	1920	1930	1991
Austria	26,153	10,442	6,801
Czechoslovakia	1,087,343	585,434	586,884
Yugoslavia	563,597	465,400	378,997
Romania	1,664,805	1,552,563	1,620,199
Soviet Union	–	–	155,711
Total	3,341,898	2,613,839	2,748,592

Hungarians fleeing from the neighboring countries did not appear in Hungary in large numbers until the late 1980s, since migration between the former socialist countries was regulated by a so-called "fraternal" cooperation. The immigration of refugees from Greece, Africa, and Central and South America was, in turn, regulated by the principle of international solidarity.[19]

Consequently, the number of immigrants was around 1,500 each year up to 1984, but by 1988 it was nearly 3,000. These figures also indicate that immigration was strictly limited by a coordinated policy toward emigrants from the former socialist countries.[20] Distrust being a fundamental feature of the regime, foreigners coming from abroad and Hungarian citizens travelling there were strictly checked. Except for those leaving or, according to Communist usage, "deserting" the country illegally, Hungarians could work or study abroad only officially, in the framework of bilateral agreements. Foreigners (Poles, Cubans or Vietnamese) could come to work or study in Hungary on a similar basis.

From the 1960s people were allowed to go on private tours to the West every third year. In the early 1980s an ever growing number of nominal marriages contributed to breaking up the system. Qualitative changes began in 1988 after four decades of controlled and restricted migration. These changes were in connection with the last phase of Ceausescu's rule in Romania, the outbreak of the war between the Croats and the Serbs in Yugoslavia, and with the collapse of the Bolshevik-type political regime in the Soviet Union.

International Migration Affecting Hungary Between 1988 and 1994

International migration as a natural process is closely related to the social, political, and economic situation both in the recipient and the releasing countries. Except for critical situations, the number of the immigrants is determined by the needs of the recipient countries.

It is not easy to find a direct relationship between refugees and immigrants, and other people applying for citizenship in a given country. In the case of Hungary the relationship is more obvious owing to the historical background outlined above.

Refugees

In 1988 more than 13,000 Romanian citizens belonging to the Hungarian nationality remained in Hungary and they were joined by even more in 1989. The political elite slowly losing power at that time was confronted by a new problem, as Hungary, following the demands of the Socialist bloc, had not joined the Geneva Convention of 1951 and the New York Protocols of 1967 following from it. However, political changes in the late 1980s made it possible to bring the legal status of refugees under control according to the international legal practice in 1989, prior to the change of regimes.[21]

As a result of the new situation, 127,108 refugees arrived in Hungary between 1988 and 1994, 99.2 per cent of whom (i.e., 126,209) came from Yugoslavia, Romania, and the Soviet Union.

Number and Rate of Refugees According to Their Citizenship and the Year of Their Arrival

Year	Total		From Romania		From the Soviet Union		From Yugoslavia		Others	
	#	%	#	%	#	%	#	%	#	%
1988	13,173	100.0	13,173	100.0	–	–	–	–	–	–
1989	17,448	100.0	17,365	99.5	50	0.3	–	–	33	0.2
1990	18,283	100.0	17,416	95.3	488	2.7	–	–	379	2.1
1991	53,359	100.0	3,728	7.0	738	1.4	48,485	90.9	408	0.8
1992	16,204	100.0	844	5.2	241	1.5	15,021	92.7	98	0.6
1993	5,366	100.0	548	10.2	168	3.1	4,593	85.6	57	1.1
1994	3,275	100.0	661	20.2	204	6.2	2,386	72.9	24	0.7
Total	127,108	100.0	53,735	42.3	1,889	1.5	70,485	55.5	999	0.8

In the above mentioned period the majority of refugees (55.4 %) came from former Yugoslavia. They were closely followed by those coming from Romania. In the first three years of the period immigrants from Romania were the most numerous. In 1989, the rate of refugees from the Soviet Union came to a mere 0.3 per cent, and even in 1990 they constituted only 2.7 per cent. Romanian citizens came in the greatest numbers in 1990, when 32.4 per cent of the total number of Romanian immigrants arrived. The distribution of refugees changed drastically in 1991 when the number of those coming from former Yugoslavia nearly reached the total number of those coming from Romania in the first three years. In the following year, the number of refugees started to decline, then decreased significantly. In 1994, Yugoslavs were no more predominant among the refugees, but their rate still amounted to 70.7 per cent. The rate of those coming from Romania was 19.6, and that of former Soviet citizens was 6 per cent in the same year.

Except for the year 1989 when 79.4 per cent of all refugees arrived illegally, most of them came to Hungary legally. Following the events in Romania in 1990, 18.6 per cent of the refugees chose illegal ways of coming, while in 1991 and 1992 the rate of illegal arrivals was less than 12 per cent. The rate of those arriving illegally was the lowest in 1993 (1.7 %), then in 1994 it increased again to 4 per cent.

In the past seven years nearly 55 per cent of the refugees—from 1991 onward the Croats and Hungarians escaping from Serbia because of the war, then the Hungarians from Croatia, and from 1995 onward the Moslims from Bosnia—were granted a temporary refugee status. Most of them did not want to settle down in Hungary, but rather to return to their homeland or to go to a third country. Procedures for permanent refugee status were launched in 4 per cent of the cases (5,610 people), as a result of which 3.2 per cent of the refugees (4,102 people) were granted the status of permanent refugee. So 41 per cent of the refugees did not apply for either permanent or temporary refugee status.

Ethnic Distribution of Refugees Coming to Hungary Between 1988 and 1994

	1988	1989	1990	1991	1992	1993	1994	1988	1989	1990	1991	1992	1993	1994
	Number							Rate (%)						
Albanian	–	–	58	90	94	28	83	–	–	0.3	0.2	0.6	0.5	2.5
Bosnian	–	–	–	1,019	7,136	382	297	–	–	0.0	1.9	44.6	7.1	8.8
Bulgarian	4	12	–	–	–	–	–	0.0	0.1	–	–	–	–	–
Gipsy	31	62	–	–	–	–	–	0.2	0.4	–	–	–	–	–
Czech	21	58	–	–	–	–	–	0.2	0.3	–	–	–	–	–
Croat	–	–	–	28,557	1,759	158	68	–	–	0.0	53.0	11.0	2.9	2.0
Polish	–	2	–	–	–	–	–	–	–	–	–	–	–	–
Hungarian	11,745	10,821	13,249	22,626	5,829	4,321	2,572	89.2	62.0	72.5	42.0	36.5	80.5	76.2
German	256	805	–	–	–	–	–	1.9	4.6	–	–	–	–	–
Russian	–	4	161	171	74	14	32	–	0.0	0.9	0.3	0.5	0.3	0.9
Romanian	1,097	5,545	1,564	930	192	112	117	8.3	31.8	8.6	1.7	1.2	2.1	3.5
Serb	–	6	–	115	109	157	79	–	–	0.0	0.2	0.7	2.9	2.3
Slovak	–	8	–	–	–	–	–	–	–	–	–	–	–	–
Ukrainian	–	–	19	91	43	6	22	–	–	0.1	0.2	0.3	0.1	0.7
Other Soviet	6	40	–	–	–	–	–	0.0	0.2	–	–	–	–	–
Other	13	85	3,232	266	753	188	105	0.1	0.5	17.7	0.5	4.7	3.5	3.1
Total	13,173	17,448	18,283	53,864	15,989	5,366	3,375	100.0	100.0	100.0	100.0	100.0	100.0	100.0

The national distribution of refugees arriving in 1988 and 1989 is special and differs from that of those coming later. In these years 99.7 per cent of the total number of refugees came from Romania. Of these people, 74 per cent declared themselves Hungarian fleeing from an ever stricter policy against the nationalities. The rate of Romanians wanting to leave Romania amounted to 21.5 per cent of the people coming from Romania, while that of Romanian citizens of German nationality to 3.4 per cent. A mere 0.3 per cent of those coming from Romania was made up by Gipsies, Bulgarians, Czechs, Poles, Serbs, and other former Soviet citizens in 1989. The rate of Hungarians fell from 89 per cent in the previous year to 62 per cent. At the same time, the rate of ethnic Romanians among the newcomers was five times higher than before, and that of ethnic Germans increased by more than two hundred per cent. The tendency was similar also in the case of other nationalities coming from Romania in smaller numbers.

Between 1990 and 1994 96,877 people got into connection with the immigration authorities in Hungary in one way or another. 50.2 per cent of them were ethnic Hungarians, 31.5 per cent were Croats, 9.1 per cent were Bosnians, and 3 per cent were Romanians. There were also Serbs, Russians, Albanians, and Ukrainians. All the rest amounted to 4.7 per cent. The rate of Hungarians was the highest in 1993 when it amounted to 80.5 per cent of all refugees coming to Hungary.

In 1994, the total number of newcomers was 3,375, 70.7 per cent of whom arrived from the former Yugoslavia. 2,016 of these people came from Rump-Yugoslavia, and only 324 persons came from Bosnia. They were followed by people coming from Romania and the Commonwealth of Independent States. In 1994 239 persons were recognized as permanent refugees, and 2,232 persons as temporary ones. Owing to various causes, 904 persons were not given protection (the application of 29 persons was rejected, and the permanent refugee status was revoked in 534 cases).

Taking the national distribution of all newcomers of 1994 into consideration, 76.2 per cent of them (and 76 per cent of those coming from former Yugoslavia, i.e., 1,813 persons) were ethnic Hungarians. The category "others" came to 9.8 per cent, the Moslims to 8.9 per cent, the Serbs to 3.1 per cent, and the Croats to 2 per cent. The Hungarians dominated also in the group of those who were granted permanent refugee status (239 persons) with 84.9 per cent. However, 97.9 per cent of those whose permanent refugee status was revoked were also Hungarians.

Turnover of the Local Organs of the Emigration and Migration Department, 1989–1994

Year	Total of newcomers #	Illegal arrivals #	Illegal arrivals %	Procedure for permanent status #	Procedure for permanent status %	Permanent status received #	Permanent status received %	Application rejected #	Application rejected %	Proceedings stopped #	Proceedings stopped %	Status withdrawn #	Status withdrawn %
1989	3,641	–	–	36	1.0	35	1.0	1	0.0	–	–	–	–
1990	15,309	316	2.1	3,520	23.0	2,561	16.7	318	2.1	548	3.6	–	–
1991	10,267	240	2.3	921	9.0	434	4.2	150	1.5	223	2.2	168	1.6
1992	5,547	117	2.1	458	8.3	472	8.5	71	1.3	58	1.0	277	5.0
1993	5,366	93	1.7	468	8.7	361	6.7	45	0.8	21	0.4	278	5.2
1994	3,375	121	3.6	207	6.1	238	7.1	29	0.9	13	0.4	534	15.8
Total	43,505	887	2.0	5,610	12.9	4.102	9.4	614	1.4	863	2.0	1.257	2.9

The table on the previous page shows the data of those whose cases were settled between 1989 and 1994 on the basis of the Geneva Convention. (When interpreting the data, I left out of consideration the 391 persons whose cases had not been settled by then out of consideration.)

During the past six years, 6,836 final decisions were passed, and subsequently 12.9 per cent of the newcomers (5,610 persons) were placed under regulations of the Geneva Convention. As a result, 402 of them were granted permanent status. The rest of the cases ended in rejection (9%), in nonsuit (13%), or in the withdrawal of the permanent refugee status (18%). The most common causes of withdrawal were the following: the person was granted Hungarian citizenship (87.9%), returned to his or her homeland (3%), disclaimed his or her status or applied for protection in his or her original country.

Number and National Distribution of Persons Getting Permanent Refugee Status, 1989–94

Citizenship	1989	1990	1991	1992	1993	1994	Total
Albanian	–	5	6	–	–	–	11
Bulgarian	1	1	–	–	–	–	2
Bosnian	–	–	–	–	2	1	3
Czechoslovakian	1	5	–	–	–	–	6
Georgian	–	–	–	–	–	20	20
Croatian	–	–	–	–	17	–	17
Yugoslavian	1	1	150	381	–	–	433
Rump-Yugoslavian	–	–	–	–	314	193	507
Russian	–	–	–	4	–	–	4
Armenian	–	–	–	3	1	8	12
Romanian	27	2,522	255	79	26	17	2,926
Soviet	5	26	23	1	–	–	55
Turkish	–	1	–	–	–	–	1
Ukrainian	–	–	–	4	1	–	6
Total	35	2,561	434	472	361	239	4,102

71.3 per cent of the permanent refugees were Romanians, and 25.4 per cent were citizens of the former Yugoslavia. The rest (3.3%) were Albanians, Bulgarians, Czechoslovakians, former Soviets (Ukrainians, Russians, Armenians, and Georgians), and Turks.

Between 1989 and 1994 614 applications were rejected. In 1991, the majority of the rejected were Romanians (74.7%), and "Soviets" (20.7%). In 1992 the rate of the Romanians was 42.3 per cent and the Yugoslavs followed with 29.6 per cent. In 1993 Rump-Yugoslavia took the lead with 73.3 per cent. In 1994 the majority of the rejected were citizens of Romania and Rump-Yugoslavia. Taken the period of four years together, 54.9 per cent of all rejected persons came from Romania, and 23.7 per cent from the former Yugoslavia.

Finally, let us examine the data referring to people receiving temporary refugee status between July, 1991 and December 31, 1994. The legal status of temporary refugees has not been regulated yet, and the procedures undertaken for them are determined by the attitude of the authorities dealing with them. Their number on December 31, 1994 was 7,738, 44 per cent of which were Hungarians. The rate of the Moslems (Bosnians) and the Croats was also significant, while that of the Serbs was only 2.8 per cent and that of the rest was 1.4 per cent. Most temporary refugees lived outside the refugee camps, 95.7 per cent being Hungarians, Moslems, and Croats. Only 9.3 per cent of the Hungarians lived in refugee camps (928 children and 2,479 adults). 65.6 per cent of those living in camps were Moslems.

53.4 per cent of the temporary refugees were men and 46.6 per cent of them were women. The rate of those older than sixty is relatively high (11 per cent). Juveniles under nineteen constituted 27.3 per cent, while people belonging to the age-group 19–59 constituted 61.7 per cent of all temporary refugees, which can be attributed to obligatory military service. In the age-group under eighteen there were only five more boys than girls (1,058 and 1,053 respectively).

Besides the data of temporary refugees on December 31, 1994, those referring to people receiving temporary asylum in 1992 (16,204 persons) have also been available for us. The most numerous group in 1992 was that of those coming from former Yugoslavia.[22]

Ethnic Distribution of People Coming From Yugoslavia and Applying for Temporary Refugee Status in 1992

Nationality	Total		Living			
			in refugee camps		with Hungarian families	
	number of people	per cent	number of people	per cent	number of people	per cnet
Hungarians	5,059	33.7	709	14.0	4,350	86.0
Croats	1,759	11.7	380	21.6	1.379	78.4
Moslems	7,136	47.5	4,616	64.7	2,520	35.3
Others	1,067	7.1	577	54.0	490	46.0
Total	15,021	100.0	6,282	41.8	8,739	58.2

Bosnian Moslems constituted the most numerous group with 47.5 per cent, they were followed by the Hungarians with 33.7 per cent, and the Croats with 11.7 per cent. Many refugees had either relatives or acquaintances in Hungary, or were in the position to pay for accomodation, for 58.2 per cent of those coming in 1992 lived outside the refugee camps. Most of these people were Hungarians (86%), while only 35.3 per cent of the Moslems could afford living with families.

The Dayton Agreement of 1995 provides for the reannexation of occupied territories, but the refugees coming from former Yugoslavian territories will have to wait before being able to return to their homes.

As a consequence of the fact that Hungary joined the Geneva Convention of 1951 in 1989 with territorial restrictions, refugees coming from countries outside Europe can apply for refugee status with the Budapest representative of the United Nations' High Commissionaire for Refugees. Since 1990, 2,183 persons turned to the Budapest office and 152 of them were granted permanent refugee status.

Immigrants

Immigration and emigration were regulated as required by international law only shortly before the change of political regimes in Hungary, in

1989.[23] Under these new circumstances 101,425 immigrants arrived from 169 countries in the period between 1988 and 1994, so their rate within the total population rose from 0.02 percent at the beginning of the period to over 1 per cent at the end of it. (In Italy the same rate changed from 0.5 per cent in 1980 to 1.5 per cent in 1990.)[24] Most immigrants arrived in 1989, and their number dropped to below 10,000 in 1993 and 1994.

Number of Immigrants to Hungary, 1988–1994

Year	Men	Women	Total
1988	5,353	5,587	10,940
1989	10,798	11,145	21,943
1990	9,739	9,218	18,957
1991	10,085	8,831	18,916
1992	7,155	6,888	14,043
1993	4,894	4,611	9,505
1994	3,899	3,222	7,121
1988–1994	51,923	49,502	101,425

The distribution and other characteristics of the immigrants are determined by the fact that 71.8 per cent of them came from Romania, 9.3 per cent from former Yugoslavia, and 8.4 per cent of them from the former Soviet Union. There were also Austrians (0.1%), Czechoslovakians (1%), and people from various other countries (9.5%) among them.

In the period 1988 to 1994 51.2 per cent of all immigrants were men and 48.8 per cent were women, which is almost the exact opposite of the distribution of men and women in Hungarian society in general (48.1% men and 51.9% women). Most of them belonged to the age-group 15–49 (70%). This age-group was the most numerous in 1991, but its share has diminished since then. However, the age-group of those above fifty increased from 9.8 per cent in 1988 to nearly 15 per cent in 1994. The participation of the age-group 0–14 dropped from 21 per cent to 10.6 per cent in the same period (see Table 1).

The change in the distribution of German immigrants is also very interesting. In 1988 and 1989 former East German citizens dominated among the immigrants, but after 1989 the number of those coming from Western

Germany became dominant. Many of them had probably been former Hungarian citizens. Immigrants also arrive continuously although not numerously from Arab countries like Algeria, Egypt, Iraq, Iran, Lebanon, etc..

Listing the countries with more than 1,000 emigrants we get the following figures: Romania 71.8%, former Yugoslavia 9.4%, the former Soviet Union 8.4%, Germany 1.8%, former Czechoslovakia 1.8%, and China 1.3%. This list reveals that similarly to refugees, immigrants also came primarily from three countries, namely, from former Yugoslavia, Romania, and the former Soviet Union. Unfortunately no separate data exists for the distribution of the families and the number of children, but the rate of men and women, and the age distribution of immigrants indicate that there must be many of them who have families and one or more children.

Since 1990, the most numerous age-group within the category 0 to 14 years of age has been children under four, indicating that the rate of younger immigrants with families has grown. This statement is supported also by the fact that within the category 15 to 49 it is the age-groups 20–24 and 25–29 that are the most numerous where young women in their most fertile years belong. The younger generations dominate among the immigrants, anyway, so their age distribution is much more favourable than that of the total population of the country. However, the rate of women lags behind that of men nearly in all age-groups, the demographic consequences of which should not be left out of consideration. The distribution of juveniles according to sex is balanced. Although there were 2 per cent more boys in 1988 and 1989, their overall rate was only 0.4 per cent more than that of the girls in the period.

The average age of those coming from Romania was the lowest among the immigrants, which was due to the fact that up to 1993 the Romanian age-group 0–14 was more numerous and the rate of those belonging to the age-group above fifty was lower. However, in 1992 the rate of girls belonging to the age-group 0–14 was lower by 2.3 per cent, while that of the boys only by 0.4 per cent. At the same time, those above fifty were 3 per cent more in 1993 and 5 per cent more in 1994. Within the age-group 15–49 most people were between 20 and 39, though the rate of those between 35 and 39 diminished after 1990. In the age-group 20–24 there are more women than men, the difference being over 5 per cent in 1994. The cause of this may be that due to freer movement and easier administrative measures the number of marriages increased between immigrants and Hungarian citizens. The rate of people of both sexes above fifty and even more above sixty was lower before 1992, but then it started to rise and in 1994 the rate of those above fifty coming from Romania was nearly 20 per cent among the total number coming from that country.

As compared to the total number of immigrants, the rate of former Soviet citizens belonging to the age-group 0–14 was lower, and that of those above fifty, and especially sixty was much higher than the average. Since many women who immigrated to Hungary earlier and were nationalized got married there, some of these elderly people may be their parents who try to escape from their sad plight in the former Soviet Union by moving to their daughters to Hungary.

Up to 1991 only few immigrants came from former Yugoslavian territories. In 1990 their number was only 160 all in all. But then their number started to increase rapidly till 1993, only to decrease by one thousand in 1994. The majority of ethnic Hungarians formerly living in Croatia migrated to Hungary during the war years. From the Voivodina the young men of military age fleeing from forced mobilization arrived first, then came the ethnic Hungarians who saw their existence as an ethnic group endangered in Yugoslavia. This was reflected also in the age distribution of the immigrants, since the rate of the age-groups 30–34 and 35–40 was 3–7 per cent higher among those coming from Rump-Yugoslavia than among those coming from Romania.

Women predominated among the Czechslovakian immigrants, they were twice as many as men (69.6%). The difference is most conspicuous in the age-group 15–49, where there are three times as many women as men. The majority belonged to the Hungarian nationality of Slovakia and got married in Hungary.

As regards foreigners of Hungarian nationality, data are available only for 1990, 1991, and 1992. During these three years, 28,698 Hungarians arrived from 38 countries as foreign citizens and spent more than a year in Hungary. This means that 57.3 per cent of all immigrants in that period were Hungarians, students included. Without students the same rate was 65.3 per cent. 62.9 per cent of those coming from Romania, 51.5 per cent of those from former Yugoslavia, and 34.7 per cent of those from the former Soviet Union belonged to the Hungarian nationality. The rate of the age-group under fourteen was conspicuously low among foreign Hungarians (1.8%), and the rate of the age-group above fifty could not be called high, either (9%). At the same time, the age-group 15–49 constituted 89.2 per cent of all immigrants of Hungarian origin. This indicates that single people and families with few children predominated in this group of immigrants. Although we do not wish to exaggerate the influence of this fact on Hungarian demographic conditions, it may effect the number of the Hungarian population negatively in the future. If the age distribution of foreigners of Hungarian origin is compared with that of the total immigrant population in the given period, it is obvious that the difference is very little in the case of people above fifty, but it is quite significant in the age-groups 0–14 and 15–49. This phenomenon can presumably be attributed to the fact that younger people are more likely to turn their backs on living in minority status than their more settled elders. It is also interesting to note that the number of Hungarians returning home from the United States, Australia, Canada, Germany, Switzerland, and Sweden is growing year by year, and the majority belongs to the age-group over fifty. Having left the country for one reason or another several decades ago, these people usually decide to return once the former political regime is over.

As Table 2 indicates, manual workers, intellectuals, and the category "others" dominated among immigrants of both sexes in the period between 1988 and 1994. With the exception of 1988, when the rate of female intellectuals exceeded that of manual laborers by 1.1 per cent, the latter category pevailed. The number of intellectuals did not fluctuate significantly. However, in 1991 42.7 per cent of the male immigrants were manual workers, and in 1992 only 34.6 per cent, which indicates that there were changes on the Hungarian labor market and unemployment was growing. The rate of men both among the intellectuals and the manual workers was higher than that of women each year in the period in question. But while among the intellectuals the difference was above 4 per cent only in 1989, it fluctuated between 19.5 and 25.9 per cent among manual workers.

There were more women among the employees (23.2–29.8%) than men (12.1–18.4%), which comes from the nature of these jobs. While the most numerous group among all male immigrants was that of the manual workers, among women it was the category "others" that dominated. The rate of old age pensioners was also rising throughout the period, especially from 1992 onward, and reached over 16 per cent in spite of the low average of the first four years.

Besides the technical intelligentsia, doctors, and secondary-school teachers, who dominated among the intellectuals, the rate of those engaged in humanities was also significant. 80 per cent of the male manual workers were industrial workers and artisans. As regards employees, women definitely dominated among office clerks and among those working in administration. There were hardly any senior clerks or persons holding leading administrative posts among them. People working in agriculture either as farmhands, skilled agricultural laborers, independent farmers or landlords were almost totally missing from among the immigrants. With the exception of skilled agricultural laborers, their rate is less than 0.5 per cent. It is again the sign of the change of political regimes in Hungary that there are former capitalists and landlords among the immigrants, even if in very small numbers.

New Citizens

The demand for regulating citizenship in Hungary arose for the first time in the early nineteenth century. A Citizenship Act was, however, passed only as late as 1879. Three more laws have been framed since then (in 1948, 1957, and 1993), but none of them is retroactive, so they have to be applied very carefully for a given period.[25]

The Citizenship Law of 1879 was amended in 1921. This modification of the law was connected with World War I when the victorious Great Powers were very liberal in matters of citizenship of the population living on the annexed territories. Point 61 of the peace treaty namely provided that Hungarians living in an annexed territory automatically lost their former citizenship on the date of the conclusion of peace (i.e., on July 26, 1921) and became citizens of the country within the boundaries of which they were forced to live. The Citizenship Law of 1921 was meant to serve keeping one's citizenship also in the disannexed territories, but the majority of Hungarians living in the successor states did not make use of their right for option, since they thought it to be against their interests. The right for option expired on July 26, 1922, but former Hungarian citizens abroad could apply for repatriation under Paragraph 24 of Act XVII of 1922 to get back their citizenship easier.

Act L of 1879 was amended for the second time by Act XIII of 1939 that provided for the loss of Hungarian citizenship in the case of those who were naturalized as citizens of other countries. This law was invalidated retroactively by Decree 11 of 1956 of the Presidential Council of the People's Republic, making invalid also the decisions passed under Act XIII of 1939 invalid. At the same time, a Cabinet Decree declared the period between September 1, 1939 and 15 September, 1947 "time in dormit", thus no one who left the country after September 1, 1929 terminated his/her Hungarian citizenship.

Citizenship law in Hungary follows the principle of *ius sanguinis*, which means that irrespective of the place of birth of a person, he or she automatically gets the citizenship of the parents. Under Citizenship Law of 1993, the child of a Hungarian citizen is a Hungarian citizen by birth. Others can receive Hungarian citizenship either by nationalization or by repatriation.

Between 1958 and 1994 74,453 persons were granted Hungarian citizenship, 30.6 per cent of whom received it prior to 1989, while 69.4 per cent after that date. People of Hungarian origin living in the neighboring countries, especially in Romania most are interested in receiving Hungarian citizenship. The change of political regimes increased the number of those claimants who had been released from or deprived of their citizenship in the past seventy years and they wish to regain it now. 14.2 per cent of the new citizens regained their citizenship by repatriation.

It is interesting to compare the four countries where the majority of the immigrants came from with those whose citizens applied for and were granted Hungarian citizenship in the greatest number. As regards immigration, Romania (71.8%), former Yugoslavia (9.4%), the former Soviet Union (8.4%), and Germany (1.8%) occupied the first four places, and in the case of citizenship the order is Romania (86.3%), the former Soviet Union (7%), former Yugoslavia (2.8%), and Bulgaria (0.9%). Romania takes the lead in both cases.

The number of those coming to Hungary from nearly eighty countries and receiving Hungarian citizenship between 1988 and 1994 was the following:

Year	Number of persons
1988–1989	1,976
1990	3,170
1991	5,893
1992	21,880
1993	11,521
1994	9,238
Total	53,678

During these years the majority of the new citizens of Hungary came from Romania each year. The rate of this group was the highest in 1992 exceeding 94 per cent. In 1993 it dropped to 86.4 per cent, and in 1994 to 67.7 per cent. Besides the countries listed above, people from former Czechoslovakia, Poland, and Germany also received Hungarian citizenship in the past five years (0.3% each), which means that while seventy-two countries of the world gave 1,075 new Hungarian citizens (with forty-two displaced persons among them), the seven above mentioned countries gave 50,627 persons, 88.1 per cent coming from Romania. The cause of this latter feature of immigration was most probably due to the great number of Hungarians living in Romania, the general economic and political situation in that country, the Romanian attitude toward minorities, the difference of the Romanian culture, traditions and customs from those of the Hungarian minority, and the attraction of those emigrating earlier. It is surprising that the number of former Yugoslavian citizens receiving Hungarian citizenship is small, 1,467 persons all in all. The cause of this may be that refugees from Croatia and Serbia of Hungarian origin lived under good circumstances before the war and they still hope to be able to return to their homeland. So they have not decided yet whether they want to remain in Hungary or not. The overwhelming majority of people coming from the former Soviet Union belong to the Hungarian nationality of Sub-Carpathia. The minority is composed of mostly Russian wives of Hungarian citizens who got married while studying or working there. Former citizens of Arab countries are also to be found among the new Hungarian citizens (0.4%).

The distribution of males and females among those receiving Hungarian citizenship since 1990 is fairly balanced. Just like in the case of immigrants, females dominate among those coming from the former Soviet Union and former Czechoslovakia, resulting from the fact that more women of Hungarian origin got married in Hungary from these countries than from Romania or former Yugoslavia. The age distribution of the new citizens is favourable, the most numerous age-groups being the one between 15 and 39, and the one between 40 and 59. The number of those above 60 is relatively insignificant. The greater number of children under fourteen makes us believe that there are several families with children among the new citizens, especially among those coming from Romania. Women in the age-groups 15–39 and 40–59 are much more numerous in the case of new citizens from the former Soviet Union and Czechoslovakia, while the distribution of men and women is more balanced in the other age-groups and in the case of people coming from Romania and Yugoslavia.

As far as the distribution of the new citizens by job status is concerned, intellectuals and manual workers constitute the most numerous categories. Their number nearly doubled each year from 1990 onward, especially if we leave out of consideration the categories "schoolchildren" and "dependants and unemployed". (The latter group consists of children under six, housewives, and old age pensioners.) Members of the medical professions are conspicuously numerous among the intellectuals.

Very few of the manual workers work in agriculture, and 50 per cent of the skilled workers are artisans. There are conspicuously few people working in the services. The category "others" consists of capitalists, landowners, independent farmers, pensioners, members of the liberal professions, and sportsmen. The overwhelming majority of the pensioners are former Romanian citizens. It is interesting to note that the number of pensioners coming from other countries has been growing in the last two years.

Women dominate among intellectuals, employees and "others", while men among manual workers. However, men are more numerous among members of the technical intelligentsia and among artists. In spite of all these, no significant difference can be established between men and women in their distribution by job status.

Table 1
Distribution of Immigrants by Age-Groups and Sex, 1988-1994

Year	1988 Men	1988 Women	1988 Total	1989 Men	1989 Women	1989 Total	1990 Men	1990 Women	1990 Total	1991 Men	1991 Women	1991 Total	1992 Men	1992 Women	1992 Total	1993 Men	1993 Women	1993 Total	1994 Men	1994 Women	1994 Total	1988-1994 Men	1988-1994 Women	1988-1994 Total
0-4	367	347	714	832	783	1615	849	772	1621	703	679	1382	552	531	1083	314	295	609	168	144	312	3785	3551	7336
5-9	404	374	778	892	872	1764	761	721	1482	575	568	1143	402	418	820	251	246	497	117	111	228	3402	3310	6712
10-14	407	388	795	984	972	1956	766	749	1515	550	510	1060	354	331	685	230	207	437	97	116	213	3388	3273	6661
0-14 total	1178	1109	2287	2708	2627	5335	2376	2242	4618	1828	1757	3585	1308	1280	2588	795	748	1543	382	371	753	10575	10134	20709
15-19	935	516	1451	1395	1023	2418	1298	1174	2472	1170	1072	2242	747	717	1464	384	382	766	414	347	761	6343	5231	11574
20-24	930	861	1791	1757	1828	3585	2252	2107	4359	2983	2595	5578	1678	1714	3392	725	892	1617	653	705	1358	10978	10702	21680
25-29	763	881	1644	1348	1404	2752	1410	1229	2639	1623	1248	2871	1346	1060	2406	728	619	1347	533	470	1003	7751	6911	14662
30-34	783	787	1570	1340	1383	2723	1241	1078	2319	1250	869	2119	893	704	1597	549	432	981	395	253	648	6451	5506	11957
35-39	585	578	1163	1139	1161	2300	985	882	1867	939	750	1689	701	544	1245	475	360	835	418	237	655	5242	4512	9754
40-49	729	607	1336	1424	1249	2673	1122	877	1999	1001	741	1742	794	615	1409	608	497	1105	574	320	894	6252	4906	11158
15-49 total	4725	4230	8955	8403	8048	16451	8308	7347	15655	8966	7275	16241	6159	5354	11513	3469	3182	6651	2987	2332	5319	43017	37768	80785
50-59	272	282	554	523	486	1009	308	323	631	298	305	603	326	385	711	332	309	641	304	242	546	2363	2332	4695
60-X	188	289	477	261	437	698	223	323	546	229	314	543	373	479	852	298	372	670	226	277	503	1798	2491	4289
50-X total	460	571	1031	784	923	1707	531	646	1177	527	619	1146	699	864	1563	630	681	1311	530	519	1049	4161	4823	8984
Total	6363	5910	12273	11895	11598	23493	11215	10235	21450	11321	9651	20972	8166	7498	15664	4894	4611	9505	3899	3222	7121	57753	52725	110478

Table 1 (continued)
Percentage of Immigrants by Age-Groups and Sex

Year	1988 Men	1988 Women	1988 Total	1989 Men	1989 Women	1989 Total	1990 Men	1990 Women	1990 Total	1991 Men	1991 Women	1991 Total	1992 Men	1992 Women	1992 Total	1993 Men	1993 Women	1993 Total	1994 Men	1994 Women	1994 Total	1988–1994 Men	1988–1994 Women	1988–1994 Total
0-4	5.8	5.9	5.8	7.0	6.8	6.9	7.6	7.5	7.6	6.2	7.0	6.6	6.8	7.1	6.9	6.4	6.4	6.4	4.3	4.5	4.4	6.6	6.7	6.6
5-9	6.3	6.3	6.3	7.5	7.5	7.5	6.8	7.0	6.9	5.1	5.9	5.5	4.9	5.6	5.2	5.1	5.3	5.2	3.0	3.4	3.2	5.9	6.3	6.1
10-14	6.4	6.6	6.5	8.3	8.4	8.3	6.8	7.3	7.1	4.9	5.3	5.1	4.3	4.4	4.4	4.7	4.5	4.6	2.5	3.6	3.0	5.9	6.2	6.0
0-14 total	18.5	18.8	18.6	22.8	22.7	22.7	21.2	21.9	21.5	16.1	18.2	17.1	16.0	17.1	16.5	16.2	16.2	16.2	9.8	11.5	10.6	18.3	19.2	18.7
15-19	14.7	8.7	11.8	11.7	8.8	10.3	11.6	11.5	11.5	10.3	11.1	10.7	9.1	9.6	9.3	7.8	8.3	8.1	10.6	10.8	10.7	11.0	9.9	10.5
20-24	14.6	14.6	14.6	14.8	15.8	15.3	20.1	20.6	20.3	26.3	26.9	26.6	20.5	22.9	21.7	14.8	19.3	17.0	16.7	21.9	19.1	19.0	20.3	19.6
25-29	12.0	14.9	13.4	11.3	12.1	11.7	12.6	12.0	12.3	14.3	12.9	13.7	16.5	14.1	15.4	14.9	13.4	14.2	13.7	14.6	14.1	13.4	13.1	13.3
30-34	12.3	13.3	12.8	11.3	11.9	11.6	11.1	10.5	10.8	11.0	9.0	10.1	10.9	9.4	10.2	11.2	9.4	10.3	10.1	7.9	9.1	11.2	10.4	10.8
35-39	9.2	9.8	9.5	9.6	10.0	9.8	8.8	8.6	8.7	8.3	7.8	8.1	8.6	7.3	7.9	9.7	7.8	8.8	10.7	7.4	9.2	9.1	8.6	8.8
40-49	11.5	10.3	10.9	12.0	10.8	11.4	10.0	8.6	9.3	8.8	7.7	8.3	9.7	8.2	9.0	12.4	10.8	11.6	14.7	9.9	12.6	10.8	9.3	10.1
15-49 total	74.3	71.6	73.0	70.6	69.4	70.0	74.1	71.8	73.0	79.2	75.4	77.4	75.4	71.4	73.5	70.9	69.0	70.0	76.6	72.4	74.7	74.5	71.6	73.1
50-59	4.3	4.8	4.5	4.4	4.2	4.3	2.7	3.2	2.9	2.6	3.2	2.9	4.0	5.1	4.5	6.8	6.7	6.7	7.8	7.5	7.7	4.1	4.4	4.2
60-X	3.0	4.9	3.9	2.2	3.8	3.0	2.0	3.2	2.5	2.0	3.3	2.6	4.6	6.4	5.4	6.1	8.1	7.0	5.8	8.6	7.1	3.1	4.7	3.9
50-X total	7.2	9.7	8.4	6.6	8.0	7.3	4.7	6.3	5.5	4.7	6.4	5.5	8.6	11.5	10.0	12.9	14.8	13.8	13.6	16.1	14.7	7.2	9.1	8.1
Total	100.0	100.0	100.0	100.0	100.0	100.0	100.0	100.0	100.0	100.0	100.0	100.0	100.0	100.0	100.0	100.0	100.0	100.0	100.0	100.0	100.0	100.0	100.0	100.0

Table 2
Distribution of Potential Immigrants by Their Job Status and Sex

Profession	1988		1989		1990		1991		1992		1993		1994		1988–1994		
	Men	Women	Men	Women	Men	Women	Men	Women	Men	Women	Men	Women	Men	Women	Men	Women	Total
Intellectuals	1124	1054	1844	1591	1657	1450	1585	1146	1240	1027	548	595	266	343	8264	7206	15470
Employees	343	1042	581	1661	346	1158	771	1534	641	1087	277	444	90	169	3049	7095	10144
Manual workers	2007	994	4257	2510	4055	2221	4302	2208	2476	1224	951	492	397	235	18445	9884	28329
Others	505	1040	1147	2285	1057	1717	1345	1796	1075	1670	343	760	205	375	5677	9643	15320
Retired	196	347	261	471	248	430	254	390	415	600	295	455	211	298	1880	2991	4871
Children under 14	1178	1109	2708	2627	2376	2242	1828	1757	1308	1280	473	433	206	165	10077	9613	19690
Students											446	527	220	295	666	822	1488
Total	5353	5586	10798	11145	9739	9218	10085	8831	7155	6888	3333	3706	1595	1880	48058	47254	95312

Table 2
(continued)
Distribution of Potential Immigrants by Their Job Status and Sex in Percentage

Profession	1988 Men	1988 Women	1989 Men	1989 Women	1990 Men	1990 Women	1991 Men	1991 Women	1992 Men	1992 Women	1993 Men	1993 Women	1994 Men	1994 Women	1988–1994 Men	1988–1994 Women	1988–1994 Total
Intellectuals	21.0	18.9	17.1	14.3	17.0	15.7	15.7	13.0	17.3	14.9	16.4	16.1	16.7	18.2	17.2	15.2	16.2
Employees	6.4	18.7	5.4	14.9	3.6	12.6	7.6	17.4	9.0	15.8	8.3	12.0	5.6	9.0	6.3	15.0	10.6
Manual workers	37.5	17.8	39.4	22.5	41.6	24.1	42.7	25.0	34.6	17.8	28.5	13.3	24.9	12.5	38.4	20.9	29.7
Others	9.4	18.6	10.6	20.5	10.9	18.6	13.3	20.3	15.0	24.2	10.3	20.5	12.9	19.9	11.8	20.4	16.1
Retired	3.7	6.2	2.4	4.2	2.5	4.7	2.5	4.4	5.8	8.7	8.9	12.3	13.2	15.9	3.9	6.3	5.1
Children under 14	22.0	19.9	25.1	23.6	24.4	24.3	18.1	19.9	18.3	18.6	14.2	11.7	12.9	8.8	21.0	20.3	20.7
Students											13.4	14.2	13.8	15.7	1.4	1.7	1.6
Total	100.0	100.0	100.0	100.0	100.0	100.0	100.0	100.0	100.0	100.0	100.0	100.0	100.0	100.0	100.0	100.0	100.0

Table 2
(continued)
Distribution of Potential Immigrants by Their Job Status and Sex
Percentage Excluding Children

Profession	1988		1989		1990		1991		1992		1993		1994		1988–1994		
	Men	Women	Men	Women	Men	Women	Men	Women	Men	Women	Men	Women	Men	Women	Men	Women	Total
Intellectuals	26.9	23.5	22.8	18.7	22.5	20.8	19.2	16.2	21.2	18.3	22.7	21.7	22.8	24.2	22.7	22.5	22.6
Employees	8.2	23.3	7.2	19.5	4.7	16.6	9.3	21.7	11.0	19.4	11.5	16.2	7.7	11.9	10.2	14.7	12.6
Manual workers	48.1	22.2	52.6	29.5	55.1	31.8	52.1	31.2	42.3	21.8	39.4	17.9	34.0	16.5	37.6	17.5	26.8
Others	12.1	23.2	14.2	26.8	14.4	24.6	16.3	25.4	18.4	29.8	14.2	27.7	17.5	26.4	15.3	27.2	21.7
Retired	4.7	7.8	3.2	5.5	3.4	6.2	3.1	5.5	7.1	10.7	12.2	16.6	18.0	21.0	14.1	18.1	16.2
Total	100.0	100.0	100.0	100.0	100.0	100.0	100.0	100.0	100.0	100.0	100.0	100.0	100.0	100.0	100.0	100.0	100.0

Notes

1. The data were made available for me by the Department for Citizenship Affairs of the Ministry of Internal Affairs, the National Data Base of the National Police Headquarters, and the Emigration and Migration Department, for which I am very grateful.
2. *Magyarország története, vol. 8, Magyarország története 1918–1945* (History of Hungary, 1918–1945), (Budapest, 1976), 360–387.
3. *Erdély története* (History of Transylvania), 3 vols. (Budapest, 1986), 1731; A. Mitrovic, *Razgranicenje Jugoslavije sa Madjarskom i Rumunijom 1919–1920* (Novi Sad, 1975).
4. *Magyar Statisztikai Zsebkönyv* (Pocketbook of Hungarian Statistics), 1940; Dezső Elekes, "Trianon mérlege" (The Balance of Trianon), *Magyar Statisztikai Szemle* (Hungarian Statistical Review), no. 4 (1938): 358–367.
5. *Magyar Tájékoztató Zsebkönyv* (Hungarian Information Pocketbook), 2nd ed. (Budapest, 1943), 157.
6. Mihály Korom, "A második bécsi döntéstől a fegyverszünetig" (From the Second Vienna Award to the Armistice Agreement), in: *Tanulmányok Erdély történetéről* (Studies on Transylvania History), (Debrecen, 1989), 178.
7. *Magyar Tájékoztató Zsebkönyv* (Hungarian Information Pocketbook), (Budapest, 1943), 50–51; *Magyar Statisztikai Évkönyv* (Hungarian Statistical Yearbook) 1942 (Budapest, 1944), 2; dr. Zoltán Fogarasi, "A népesség anyanyelvi, nemzetiségi és vallási megoszlása törvényhatóságonkint 1941-ben" (Distribution of the Hungarian Population in 1941 According to Mother Tongue, Nationality, and Religion), in: *Magyar Statisztikai Szemle* (Hungarian Statistical Review), nos. 1–3 (1944): 1–20.
8. *Bajtársak a bajban. Lengyel menekültek Magyarországon 1939–1945* (Comrades in Need. Polish Refugees in Hungary, 1939–1945), (Budapest, 1985), 639.
9. According to Veesenmayer's report 437,402 people were carried away before July 9, 1944. See *Magyarország története*, vol. 8, *Magyarország története 1918–1945* (Budapest, 1976), 1162. (See also note 2.)
10. Sándor Balogh, Sándor Jakab, *The History of Hungary After the Second World War 1944–1980* (Budapest, 1986), 270.
11. "Kárpátalja csatlakozik a Szovjetúnióhoz" (Sub-Carpathia Joins the Soviet Union), *Új Szó* (New Word), 19 May 1945: 1. When the Soviet Union distintegrated on 21 December, 1991, this territory became part of Ukraine.
12. See for example Kálmán Janics, *A hontalanság évei* (Years in Exile), (Bern, 1980), 225–249.
13. Lajos Für, *Mennyi a sok sírkereszt? Magyarország embervesztesége a második világháborúban* (Hungarian Graves. Hungarian Loss in Lives in World

War II), 2nd ed. (Budapest: Püski, 1989), 35; Tamás Stark, *Magyarország második világháborús embervesztesége* (The Human Loss of Hungary in World War II), (Budapest, 1989), 71; A. Ferenc Szabó, "Az apokalipszis mérlege" (The Balance of the Apocalypse), *Valóság* (Reality), no. 7 (1989): 67–77.

14. Gyula Borbándi, *A magyar emigráció életrajza 1945–1985* (Hungarian Emigration, 1945–1985), (Bern, 1985), 11–12. The UN Organization for Refugees registered more than 62,000 Hungarian refugees.
15. *Rendőrségi Közlöny* (Police Bulletin), 15 November 1947: 625–626; Ágnes Tóth, "Bibó István memorandumai a magyarországi német lakosság kitelepítésével kapcsolatban" (Memoranda by István Bibó in Connection with the Removal of the German Population of Hungary), in: Bács-Kiskun Megye múltjából (On the History of County Bács-Kiskun), 11 (Kecskemét, 1992), 330–382; Kálmán Janics, op. cit., 251–277; A. Ferenc Szabó, "Völgységi metamorfózis" (Metamorphosis in the Völgység Region), *Valóság*, no. 10 (1993):50.
16. Nations Unies Comite executiv de l,UNREF A/AC. 79/73. 8. mai 1957; *Statisztikai Szemle* (Statistical Review), nos. 10 and 11 (1990): 987–1002.
17. Mária L. Rédei, "A nemzetközi népességmozgás negyven éve Magyarországon" (Forty Years of International Migration in Hungary), in: Pál Tamás, András Inotai, eds., *Új exodus* (A New Exodus), (Budapest, 1993).
18. *Erdély története* (The History of Transylvania), (Budapest, 1986) 3: 1739; *Magyar Statisztikai Zsebkönyv* (Pocketbook of Hungarian Statistics) *1940* (Budapest, 1940), 22; Dezső Elekes, "Trianon mérlege" (The Balance of Trianon), in: *Magyar Statisztikai Szemle* (Hungarian Statistical Review), no. 4 (1938): 358–367.
The total number of the Hungarian population was 7,986,875 in 1920, 8,685,109 in 1930, and 10,354,842 in 1991.
19. *Hetven év. A romániai magyarság története 1919–1989* (Seventy Years. History of Hungarians in Rumania, 1919–1989), (Budapest, 1990), 113–115.
20. See "1982. évi 19. törvényerejű rendelet a külföldiek magyarországi tartózkodásáról" (Executive Order 19 of 1982 on the Stay of Foreigners in Hungary), in: *Törvényerejű rendeletek 1982* (Executive Orders 1982) 3:374; "7/1982. (VIII. 26.) BM számú rendelet a külföldiek magyarországi tartózkodásáról szóló 1982. évi 19. számú törvényerejű rendelet végrehajtásáról" (Decree no. 7/1982 [26 August] of the Ministry of Internal Affairs on the execution of Executive Order 19 of 1982), in: *A belügyminiszter rendeletei 1982* (Decrees of the Minister of the Interior), 128–129.
21. "1989. évi törvényerejű rendelet a menekültként elismert személyek jogállásáról" (Executive Order from 1989 on the Legal Status of Persons Recognized as Refugees), in: *Törvényerejű rendeletek 1989*, 3: 489; "101/1989 (IX. 28) MT rendelet a menekültként való elismerésről" (Cabinet Decree

101/1989 [28 September] on Recognizing People as Refugees), in: *Miniszter-tanácsi rendeletek 1989,* 11:27–33.
22. *Demográfia,* no. 1 (1993): 49–50.
23. "1989. évi XXIX. törvény a ki- és bevándorlásról" (Act XXIX of 1989 on Emigration and Immigration), in: *Tanácsok Közlönye* (Council Bulletin) 26: 657–659; "12/1989. (XII. 29.) BM rendelet a ki- és bevándorlásról szóló 1989. évi XXIX. törvény végrehajtásáról" (Decree of the Ministry of Internal Affairs 12/1989 [29 December] on the Execution of Act XXIX of 1989 on Emigration and Immigration), in: *A belügyminiszter rendeletei 1989* (Decrees of the Minister of the Interior), 689–694.
24. International Migration: Facts, Figures, Policies, Migration. The OECD Observer 176. June/July 1992: 19.
25. Dr. Mária Parragi, dr. Mária Ugróczky, eds., *A magyar állampolgárságra vonatkozó szabályok* (Rules Regulating Hungarian Citizenship), vols. 2 (Szemimpex Kiadó, 1993), 248–249.

László Hablicsek

Demographic Transitions and the Change of Regimes in Hungary

Introduction

It is certain that the changes taking place in Hungary between 1990 and 1995 indicate the beginning of a new era in Hungarian demography. On the basis of the data, the first period of economic transformation involves one-sided demographic changes that may contribute to the aggravation of the demographic problems of the country.

Yet it is important to emphasize that the same demographic processes are at work in the present economic system in the process of transformation as in the previous "socialist market economy". Even though the general demographic goals are the same, the factors influencing the behaviour of the population and government policy have undergone significant changes.

In the short period since the beginning of the new era significant changes can be observed in the annual demographic data indicating that new behaviour patterns are taking shape. In order to avoid making major mistakes in assessing those developments it is important to adopt new demographic research techniques. Only then is it possible to offer a complex survey of the demographic changes inherited from the past, or effected by the introduction of economic and political reforms. It is imperative to understand the full weight and size of the expectable problems, and to explore their causes. Just as in other East Central European countries, the aim is not only to gain information about the general demographic behaviour of the population, but to develop data bases and methods with which problems of demographic nature can be explained and solved.

In the present article we wish to introduce Hungarian demographic processes that might serve as a starting point for the reinterpretation of the evolution under radically changing circumstances.

The pivotal question of this approach could be the relationship of the present economic transition to the long-term demographic trends. These trends characterizing the changes in the population in the nineteenth and twentieth centuries are called "demographic transitions."[1] If the effects of the present economic and political changes on the population are interpreted in this framework, we might point out demographic changes it has engendered and how far it is continuing or accelerating processes begun in the previous period; what did it end and what new elements did it create?

Demographic transitions and the transitionary stages within them are large-scale changes of behavioral patterns related to long-term economic and social changes.[2] It takes evidently a longer period for the new conditions or patterns to prevail, so the demographic consequences of the present change of regimes in Hungary cannot be measured fully as yet.

Demographic Transitions

The nineteenth and the twentieth centuries witnessed a turn in demographic trends first in the developed countries and later also in the so-called developing countries. The high level of both mortality and fertility dropped drastically in a relatively very short time. This change came to be called the first demographic transition and spanned the period between the traditional (agrarian) society of the late nineteenth century and the creation of modern industrial society in the late 1960s.

The first demographic transition had four distinct stages.[3] Demographic conditions in the first stage are characteristic of a traditional society turning industrial. Families have many children, the average life expectancy is low, and the population is very young. Transition starts in the second stage with a rapid decrease of mortality. The number of children also changes, but less drastically. These involve the increase of the population, and a demographic boom takes place. In the third stage, the decrease of mortality is coupled with a radical decrease also in the number of children, so the increase of the population slows down, and the population gets increasingly older.

The fourth stage, that is, the demographic pattern characteristic of a developed industrial society takes shape at the end of the first demographic transition. This is a stable or stationary stage,[4] where there are few children in a family, but just enough to ensure the simple reproduction of the population. Mortality is very low, the population is old and relatively constant in numbers according to the classical theory of the first demographic transition.

However, toward the end of the twentieth century, a new large-scale decline of fertility took place beginning with the 1970s in most western countries. By the 1980s the number of births decreased drastically even in the South European countries with a traditionally high fertility. Expert forecasts speak of a possible population decline of 20 to 30 per cent within 30 to 50 years.

What are the actual prospects following the first demographic transition? Will the population really decrease by half? Will the countries consist mostly of elderly people and face immense problems of supporting the old? The experiences of the developed countries ahead of us in this process cannot give full answers to these questions, but the outlines of the beginning of a new era, i.e., the second demographic transition[5] can be observed there. The number of the children is very low, life expectancy is growing, societies are getting older, the number of the population stagnates or decreases, and there is an intensive immigration to counterbalance the decrease.

Demographers dealing with the period of the second demographic transition give the following criteria of the process:[6]

- Fertility decreases to a degree that, with all the low level of mortality, it no longer ensures the simple reproduction of the population (with the size of the population unchanged), which will sooner or later lead to the large-scale decrease of the population.
- Divorce gets increasingly frequent, producing whole generations of divorced people.
- The number of marriages decreases significantly, both first marriages and remarriages. The rate of second marriages is also increasing.
- Unmarried cohabitation is getting increasingly frequent, and couples often do not make a match of it even when children are born. The number of children born out of wedlock is increasing significantly.
- There is a significant increase in one person households. The traditional large families and even nuclear families have ceased to be the basic family type.

- Childlessness is spreading both among married couples and couples living in concubinage.
- The rate of one-parent families is growing significantly.

The above symptoms of the second demographic transition are often attributed to the spread of individualism and egoism in general and are considered unfavorable from social development aspect by most scholars. At the same time the symptoms are recognized as relevant, and seem to become general and irreversible.

The second demographic transition has proved to be an applicable theoretical framework from several aspects. Such is, for example, its global aspect, i.e., the fact that parallel with the second demographic transition in this part of the world there are countries, much more populous than those in Europe, where the first demographic transition is in full progress with its strong population boom, exercising a demographic pressure globally. Migration from the overpopulated and underdeveloped territories and its consequences point to a global second demographic transition.[7]

Another aspect of the transition is the anxiety about the demographic processes in Europe, namely that low fertility will soon lead to the drastic decrease of the population.

However, when offering a prognosis, migration has to be considered, though it might fluctuate significantly. It is, however, certain that migration will play an outstanding role in the history of the second demographic transition. At the same time, the original population of the developed countries is decreasing and aging in general, causing great social tension. This tension may even undermine the whole system of the welfare state and society becomes unmanageable with the present means and methods.[8]

A serious aging of the population[9] is one of the important characteristics of the second demographic transition. Aging in itself, i.e., increasing life expectancy, is obviously a positive process. It is not aging itself, but the excessive aging of a society and the uneven age distribution that might cause serious demographic, social, and economic problems in the near future as consequences of the second demographic transition.[10]

The First Demographic Transition in Hungary[11]

In Hungary the first demographic transition lasted approximately from the last third of the past century to the 1960s. During this period, the number of children and the mortality rate fell to one third of the earlier level, doubling the population of the country (calculated for the present territory of the state). Since the 1960s, Hungary has been in a transitionary stage from a demographic point of view.

Mortality

Mortality can best be characterized by average life expectancy at birth. This index does not simply show the expected age of the people at death, but also the level of development and modernization in a given society.

During the years of the first demographic transition, life expectancy at birth rose in Hungary from below thirty to seventy, which is due to the decrease of age-specific mortality that had manifested mainly in infant mortality and the mortality of children and young adults, i.e., of the age groups 0–14 and 15–39.

There are distinct stages in the change of mortality patterns. In the first stage, i.e., up to 1920, the decrease of overall mortality was primarily influenced by the change in infant mortality. The next period between 1921–1939 brought an increasing difference between the rate of decrease in the mortality of the age group 0–4 and that of older children to the advantage of the latter. The period between 1945 and 1960 can be considered that of the fastest transformation of mortality conditions in Hungary, bringing a general decrease in mortality in nearly all age groups.

This was the end of the process bringing about modern mortality conditions in Hungary, and the recent stage has brought a unique phenomenon in international comparison: decrease has turned into a constant increase, and the tendencies of male and female mortality started to differ significantly.

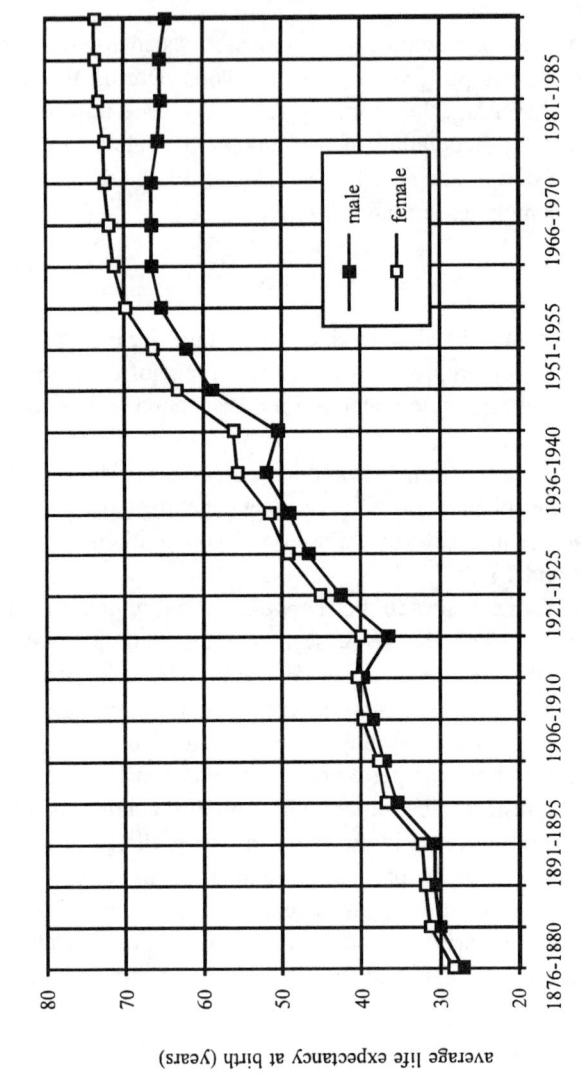

Mortality started to increase in several male age groups in the second half of the 1960s. The increase affected the whole male population between 25 and 75, and the deterioration was so serious that male mortality in the age group 45–60 fell back to the level of the 1920s.

Certain population groups are in an even worse situation (single men, people living in the Eastern part of Hungary, those on a low level of education, and manual workers). Their mortality rates equal the ones registered one hundred years ago.

Female mortality has been somewhat different from male mortality in the last thirty years. Female life expectancy has increased by a moderate two or three years, so their life expectancy surplus has grown from five years to ten in contrast with men, though middle-aged female mortality has also increased slightly.

Fertility

The changing number of children is characterized by gross and net reproduction rates, both referring to girls. The former indicates the number of daughters born to an average woman in her productive years, while the latter expresses the rate of a generation of mothers and daughters with the different mortality rates used as corrective measures.

The chart representing the changing number of children on a cross-sectional basis shows that with all the fluctuations, fertility was continuously decreasing in the period of the first demographic transition, even to a greater degree than mortality did. Fluctuation can be attributed to the two world wars, to the economic crises, and to the changing population policies.

The first major wave-trough occurred during the First World War, when a number of children corresponding to at least two cohorts were missing and were only partially replaced by the peak after the war. Families had few children also during the economic crisis of the 1930s and the Second World War, but subsequently fertility rose again and reached its peak in the first half of the 1950s, mostly owing to the suppression of induced abortion. Another decline followed in the early 1960s, to be followed by the rise of the number of children in the second half of the decade, reaching a peak in the mid-1970s as a consequence of the child-care and benefit system, introduced at that time in Hungary. Another decline followed in the 1980s.

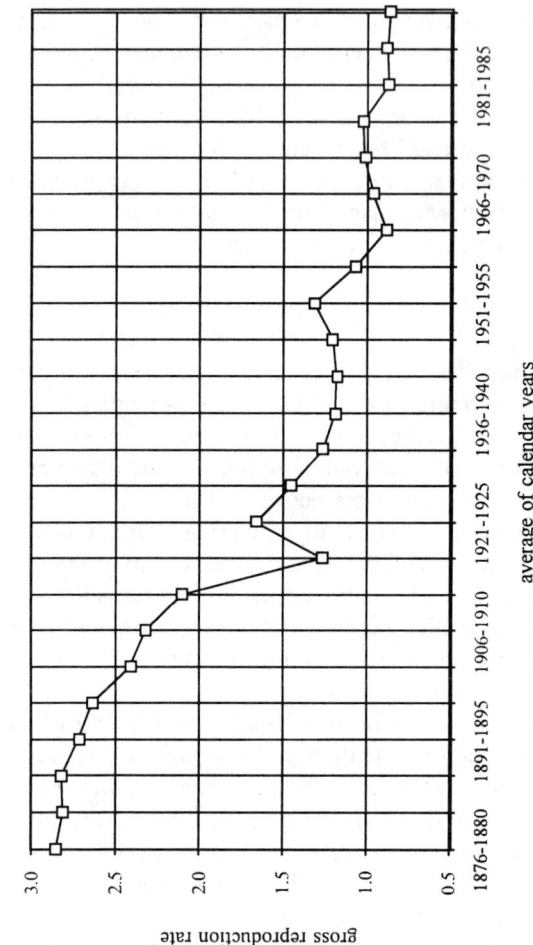

Fluctuation of the Gross Reproduction Rate, 1876-1993 (average of periods)

Examining the actual number of children of the individual cohorts instead of the cross sectional data, the fertility of the cohorts was not really influenced by the varying circumstances through the years and the above fluctuation cannot be observed, either.

This is the point where the essence of the recent changes can be found: the number of the children in a cohort group can be related to the average circumstances of a cycle of about twenty-five years. These circumstances were gradually deteriorating and becoming increasingly unfavourable as regards children. It was not only the actual number of children that was changing, but also the patterns of fertility. The subsequent cohorts kept postponing the birth of the second and the other children, and finally they postponed most of them forever. So the essence of the recent changes is that this change of patterns suddenly includes the birth of the first and the second child.

However, with the Hungarian cohorts beginning their procreative age after 1960 there was a stagnation first, but finally a significant rise occurred in the number of children, which contravenes the statement that family and child care benefits, i.e., the efforts of population policy had no real influence in Hungary.

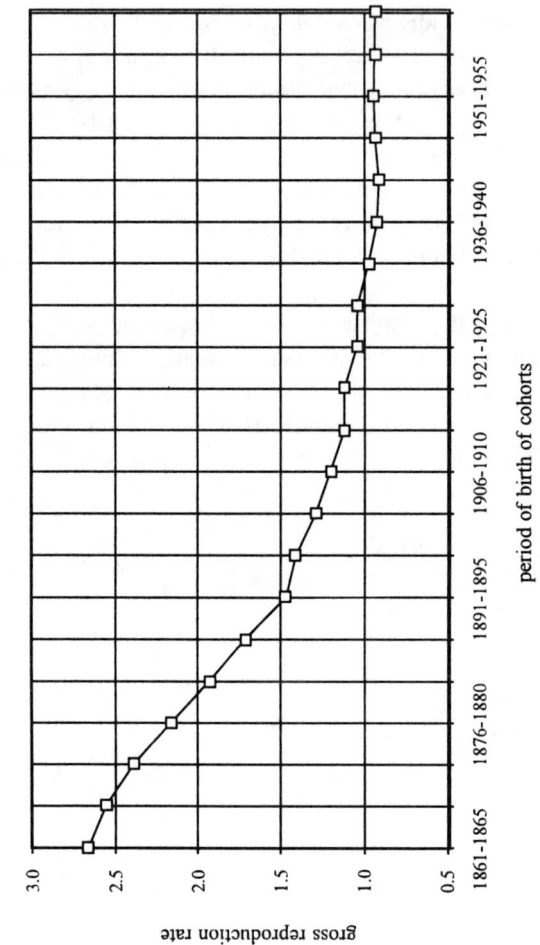

Brutto Reproduction Rate in the Individual Cohorts of Women (Completed with Estimations as to Still Productive Cohorts)

Number of Population and Age Distribution

As a combined result of the changes in fertility and mortality, the population of Hungary moderately grew during the years of the first demographic transition. This is a characteristic feature of Hungarian demographic development in international comparison. The growth was at first relatively fast, but later it slowed down. The rate of growth started to fall from 1961, and from 1981 onward the population started to diminish on the long run.

As regards the age distribution of the population, aging is significant, i.e., the older age-groups have a greater weight within the population, and the number of people belonging to the individual age-groups fluctuates significantly.

The number of those belonging to the age-group 0–14 hardly changed during the one hundred years of transition, while that of those in their working age doubled, and the number of the elderly (60 and above) increased to a greater degree than that of the whole population. So there were five times as many elderly people in Hungary in 1990 than in 1881 (on the present Hungarian territory).

International migration also belongs to the factors of the first demographic transition, although it is not widely examined. The balance of the process has influenced Hungarian demographic developments, reproduction, and the fact that the decrease of the population started so early.

The full consequences of outer migration and the basic demographic changes in the century of transition can be illustrated by models. These calculations reveal that the first demographic transition in Hungary was inevitably characterized by the parallel decrease of the number of children and mortality, and by the fact that even these two factors did not bring about either a significant increase or a decrease of the population as compared to the demographic situation the population would have produced without these changes. Migration is an exception, since it reduced the Hungarian population by nearly one million people (by nearly ten per cent) during the transition.

The decrease of the population from 1981 onwards was not an unexpected feature of the first demographic transition. Large-scale decrease would, however, be a sign of a change of trends, signifying serious troubles in reproduction.

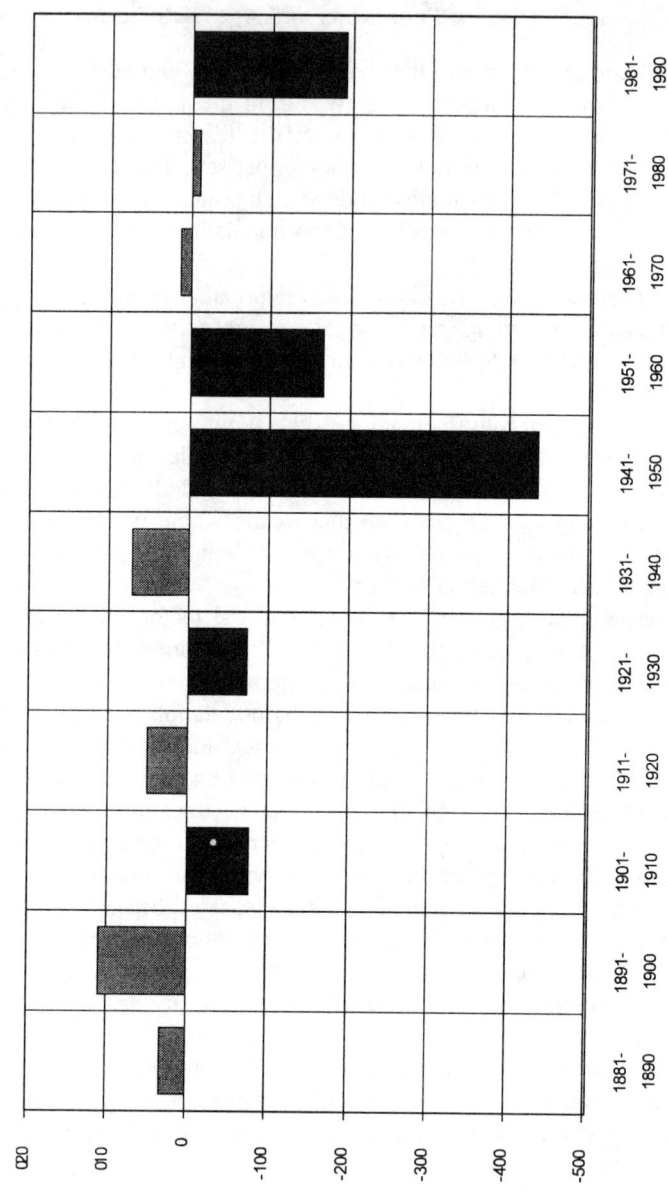

Demographic Changes Between 1970 and 1990. On the Threshold of a New Era

Between 1970 and 1990 there was a very special demographic period in Hungary. The events concentrating in these decades can probably be called a "struggle" between the first and the second demographic transitions. In the previous chapter we have described the characteristics of this period, so this time we concentrate on the trends only.

The number of children fluctuated with a peak in the seventies and a decline in the eighties. There was a period when simple reproduction seemed attainable, but it was followed by years of low fertility.

Female mortality decreased moderately, while there was just an opposite tendency for men that manifested in a shorter male life expectancy as early as the 1980s. At the same time, mortality rates for the overall population stagnated, suggesting the possibility of stabilizing population growth.

The population was still increasing in the 1970s, though very slowly, but in the early 1980s it started to decrease, which was unparalleled at that time throughout the world. It had the effect of cold water on politicians concerned with demographic processes, and refuted all efforts and theories of stabilization. In the 1980s there was one more effort to postpone the evolution of a new era in demography by political means (i.e., by the introduction of child-care fee that amounted to 75 per cent of the mothers previous income), the result of which could be seen in the second half of the 1980s and in 1990 and 1991, when the number of children was somewhat higher again.

Meanwhile, significant changes took place in the age distribution of both the society and the families that finally outdid the stimulative effect of programs encouraging couples to greater fertility, and prepared the changes of behavior characteristics from the 1990s.

Quick aging of the population played a favourable role in this process with the parallelly growing need for the extension of provisions for the retired. The number of the elderly nearly doubled between 1949 and 1990, that of the retired amounted to twenty per cent of the population in the first half of the 1980s, and by 1990 it was nearly twenty-five per cent.

At the same time, the symptoms of the second demographic transition gradually appeared as regards the families. The number of marriages decreased, while that of divorces increased greatly, and the number of children planned by the new couples was also decreasing. Nuclear families were becoming more and more general, and the number of one-person households started to increase rapidly.

By the end of the 1980s the question was when and how much these changes would become general. Both demographers and politicians hoped that the change of political regimes and the subsequent economic changes would moderate and slow down the evolving demographic "crisis" indicating the arrival of the second demographic transition in Hungary.

Estimated Demographic Changes and Their Consequences in Hungary After 1990

Right after the change of political regimes, in 1990 and 1991, demographical data reflected stability in the demographic processes. Fertility increased a bit, life expectancy stagnated, and the decrease of the number of the population slowed down, owing partly to immigration. It was the data for 1992 that first showed that the change of regimes exercised a significant unfavourable impact on demographic behaviour. It was this phenomenon that the latest estimations of 1993 and 1995 reacted upon.

Demographic Estimations up to 2020[12]

In 1992 and in the subsequent years demographic processes took a much more unfavourable turn than expected, since the number of both live births and deaths, as well as the natural decrease of the population reached their negative peaks.

According to the census of 1980, the population of Hungary was 10,712,000 while in 1990 it was only 10,375,000. Between 1990 and 1995 the population continued to decrease, and on January 1, 1995 it was 10,240,000. According to the latest estimations, by 2000 it may decrease by another 100,000–150,000, by 2010 by 400,000–600,000, and by 2020 by 900,000–1,200,000. Accordingly, in 2020 the population of Hungary may amount to 9,500,000 at best. All these indicate, that in case of the continuation of the demographic tendencies of the last five years, the decrease of the population will become increasingly irreversible.

Demographic Transitions and the Change of Regimes in Hungary 185

The Number of the Population According to Years of Age in 1993 and in 2020 (as calculated on the basis of the data from 1993)

Modelling can prove that this perspective originates from two causes. First, fertility characteristics in Hungary have become increasingly similar to those in Western Europe. Secondly, as regards mortality, Hungary is sharply deviating from these countries. Considering that a few years are not enough to draw demographic conclusions from, it is important to call attention to the fact that these tendencies are the signs of a special demographic transition in Hungary: Western fertility is combined here with Eastern mortality, producing critical conditions. One might say that the phenomena of the period 1992 to 1995 reveal divergent tendencies in the demographic attitude of the people.

The age distribution of the population reflects the fluctuation in the number of births (and that of deaths, though to a smaller degree). The main characteristic features of the changes in the age distribution are the advance of the diminishing younger generation, and the aging and decrease of the former "great" generations, mainly as regards men.

Should the characteristics of mortality valid for 1992 remain unchanged, a mere 45 per cent of the population of today will live in 2020. Men have still worse prospects with only 41 per cent. More than 90 per cent of those who were ten in early 1990 can hope to live in 2020, but only three quarters of those who were twenty-five, half of those who were forty, and hardly 10 per cent of those who were fifty-five at that time.

These survival rates are very low compared to the male population in other European countries. In Austria that represents an average in Europe, 98 per cent of those who are ten today, 92 per cent of those who are twenty-five, and more than 70 per cent of those who are forty will live to see the year 2020. These figures may explain the prevailing pessimism in Hungary, since a very significant portion of the population feels planning for more than twenty years totally useless.

The average number of children necessary to prevent the further decrease of the population with the given mortality rates can also be estimated.

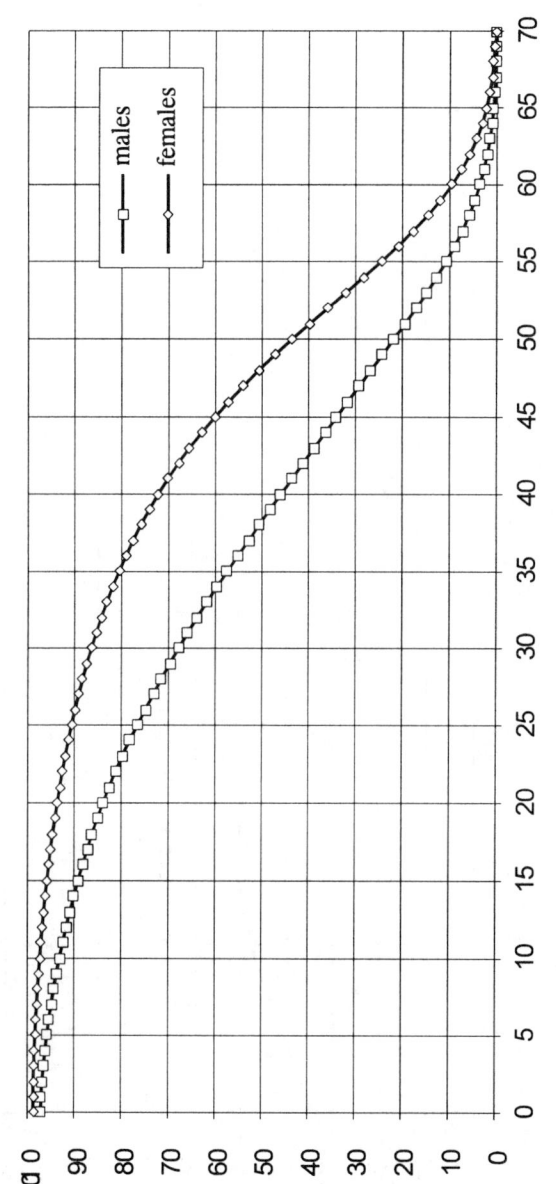

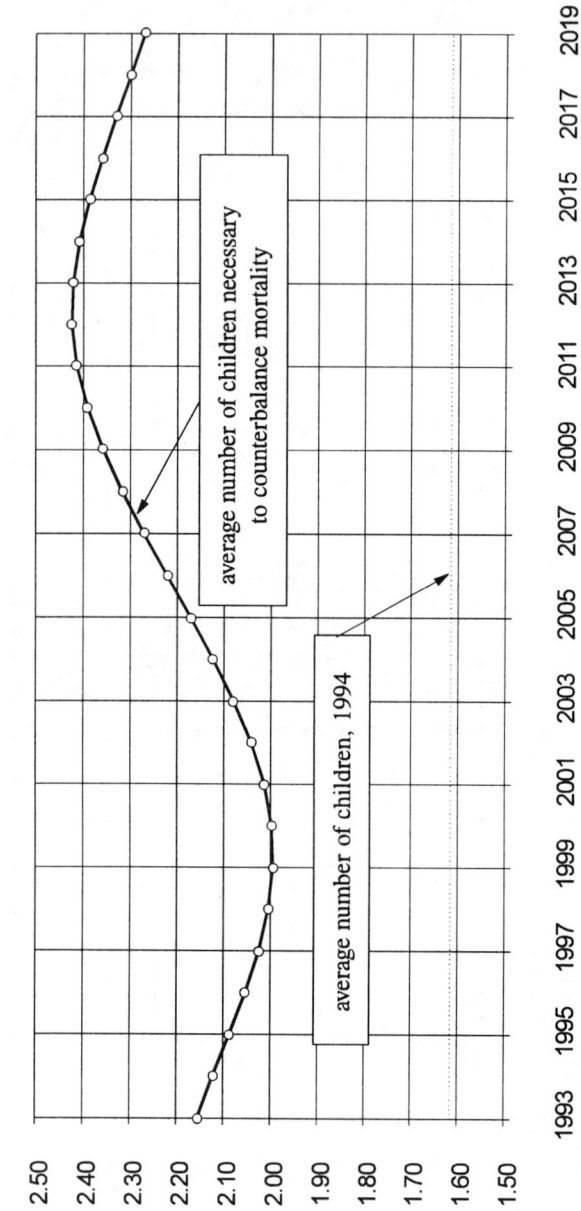

It can, therefore, be concluded that the average number of children should have been 0.5 higher in 1994, which means that every second woman should have had one more child for the number of children to reach the number of deaths in that year. In 1999 the average number of children necessary for that purpose would be around two, even if one takes into consideration that it will be the numerous generations of women born in the 1970s that will give birth then. In the first half of the 2010s the newborn babies of today will start families, and at least 80 per cent of them should have one more child than their parents had to create the balance.

The Future of the Families and the Households[13]

One of the most emphatic elements of the second demographic transition is the change in the role of the family, the loss of some of its functions, and the fact that it is increasingly dysfunctional. Since children are born mostly in legal or common law families, it seems to be obvious that advanced family policy should promote the birth of more children. Family policies should ensure proper conditions for children, as it is usual in Western Europe.

This is why the family and household estimates are becoming increasingly important. They reveal the prospective distribution of the population according to family status, types of family, and types of household. The latest estimates for the period 1993 to 2010 are based on the first significant changes in connection with the families after the change of political regimes. The essence of these new pieces of information is that the role of the traditional families and households based on marriage and kinship is rapidly diminishing, and the rate of extramarital relationships and single households is increasing.

The decrease of the number of married couples is likely to continue in the future, too, and this is going to be one of the most remarkable tendencies in the next 10–15 years. Their number is likely to decrease to a greater rate than that of the overall population. The number of the single and the divorced, and to a smaller degree also that of the widows and widowers, will continue to grow, while that of the whole population will continue to fall. The rate of married women above 15 will fall to 51 per cent, and that of married men to 57 per cent.

It is not only that the number of single people will grow, but also this demographic group will be an increasingly closed one, and many will remain single for the rest of their lives, like the majority of widows or widowers. Aging of the single and divorced population will also be remarkable. The number of the divorced in the age-group 40–59 will be 20 per cent higher, that of those in the age-group above 60 by 70 per cent. The number of single people in the age-group 40–59 will nearly double by 2010.

The above phenomena are remarkable also because they contribute to a higher male mortality. There are, namely, great differences between the mortality of married and unmarried people. A married man of forty can expect to live 7 to 9 years longer than his single, divorced or widowed counterparts. Married women of forty live 6 years longer than married men of the same age, and 15–17 years longer than unmarried men. Estimations say that the mortality rate of unmarried people will grow from 55 per cent in 1992 to 61 per cent in 2010 as compared to the overall number of deaths.

Another significant tendency is the prospective decrease of the number of families, and the families of married couples within this group. In 1990 there were 2,900,000 families in Hungary (with 2,300,000 married couples), but their number is likely to decrease by 300,000, i.e., by 10 per cent to 2,600,000 by 2010. However, the number of families with a married couple within will decrease by 330,000, i.e., by 14 per cent. The growth of the rate of unmarried cohabitation is likely to become a long-term tendency. At the same time, the number of one-parent families is likely to stagnate on the present level of half a million.

The number of people living in families (husbands, wives, partners, parents, and unmarried children) was 8,400,000 in 1990, i.e., 81 per cent of the total population. In 2010 this number is likely to fall to 7,700,000, i.e., to 78 per cent of the prospective population.

The significant decrease of the number of households based on families is going to be another new tendency in the given period. Their number is likely to fall from 2,800,000 in 1990 to 2,500,000 in 2010, which is lower than in the early 1960s. The number of one-family households is likely to decrease most, while the rate of hoseholds with several families (3–4 per cent) is likely to remain unchanged. The number of those living in family-households is going to decrease by 700,000 people, which comes nearly exclusively from the decrease of the number of family members.

Demographic Transitions and the Change of Regimes in Hungary 191

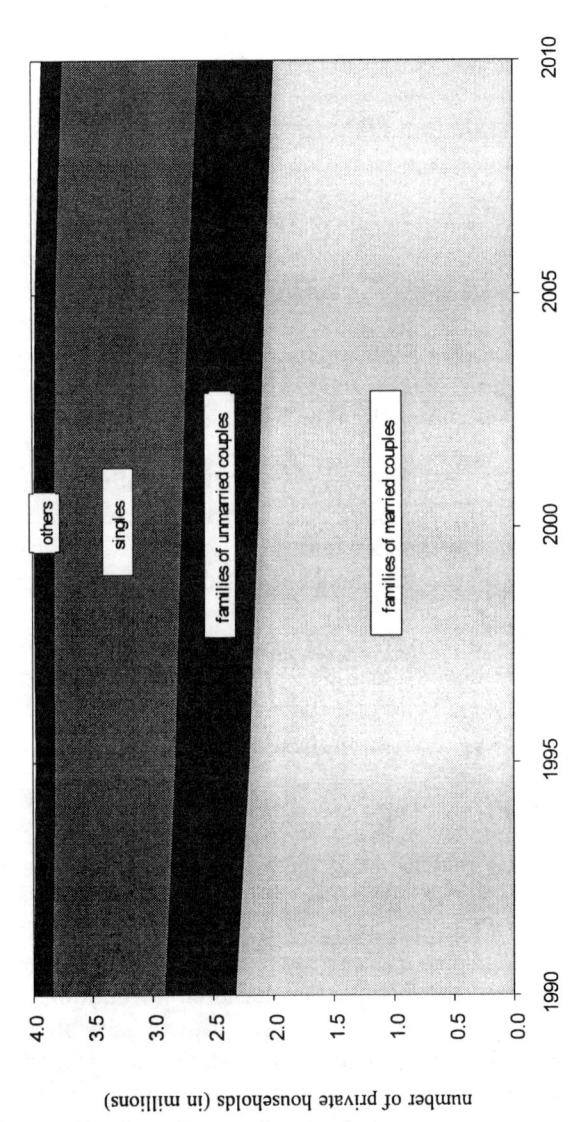

Prospective Changes in the Number of Private Households According to Main Types in Hungary, 1990-2010

The total number of private households was 3,900,000 in 1990 and will probably remain near that number also in 2010, since the decrease of the family-households will probably be counterbalanced by the growth of the number of the single households. The number of lone people will rise from 950,000 in 1990 to at least 1,140,000 in 2010. The number of households made up of relatives or other persons is also growing, so the rate of households based on relationships other than families will probably grow from 28 per cent to 34 per cent within the total number of households.

It seems that in Hungary the general demographic changes (decrease and aging of the population) take place parallelly with the reduction of the number of the traditional frameworks of reproduction (married couples, families of married couples, and family households). At the same time, the number and rate of formations unable or only partly able to ensure reproduction (unmarried couples, families without a married couple, and singles) increase. These changes will lead to a further decline in the number of children, to higher mortality, and a long-term decrease of the number of the population.

Estimations call attention to the imperative need for the introduction of a new family policy supporting long-lasting partnerships and families with children, whether they are born in or out of wedlock. Recent views that modern market economy should be based exclusively on the individuals and not on co-operating family members should be reconsidered. Special attention should be paid to the generations born in the 1970s when they start families and have children, since they constitute the last cohort in the Hungarian population that could moderate or even stop the further reduction of the population as an inner resource.

The Present and Future of Economic Activity[14]

Several experts maintain that the large-scale transformation of economic activity and inactivity during the years of the change of political regimes explains most demographic changes in the country. The previous politico-economic regime was characterised by full employment including also women, general and free public education, low retiring age, and full provision for the elderly as the statistical characteristics of economic activity and inactivity. In the first phase of the change of regimes it was employment that changed most. Open unemployment appeared, the number of the economically active population decreased, and the number of the inactive people in their working age who do not fall in any category also increased. The number of the retired grew rapidly. All the above gave rise to a significant increase of taxes and contributions, which contributed to the decrease of the actual income of the population simultaneously with the fall of the real wages due to economic problems.

In early 1994, for every one hundred people in their working age there were sixty-nine people not in their working age, for every one hundred economically active persons there were one hundred and thirty non-active, and for every one hundred earners there were one hundred and ninety-two non-earners in Hungary. The rate of unemployment was around fourteen per cent, taking only the registered unemployed people into account.

Estimations of economic activity seem to be fairly optimistic, saying that by 2000 the number of earners will rise from the present level (fifty per cent of those in their working age), and the number of the unemployed will decrease. Owing to demographic factors, the number of children in their school age will also decrease, while the number of the retired and the other economically non-active categories will continue to rise.

As we near 2010, when the populous generations born in the 1950s grow old, an even more numerous stratum will join the age-group of the elderly and the number of people in their working age will decrease even stronger.

The extension of school-age may slacken the pace of the decrease of the number of school-children, but it cannot prevent it after all. Between 2000 and 2010 there will be 90,000 to 100,000 fewer children learning at schools than today.

These estimations suggest that there is a great need for better economic and social conditions to promote a significant increase in the number of the employed. It seems to be imperative that the further decrease of the number of earners should be stopped, or else the rate of earners and dependants will be fatally distorted, which may contribute to a nihilistic attitude which prefers unemployment and inactivity to activity. The rate of the economically active should be increased, and that of the retired and other economically inactive categories lowered.

Estimations also suggest that unemployment is not the only problem, and probably not the most serious one concerning economic activity in Hungary, however important it may seem today. The whole system of activity and inactivity is endangered, owing to the shift of inner proportions and the burdens of the earners. Consequently, productivity and the value system may dangerously change as well in the near future.

Estimated Number of the Population in Categories According to Sex, Age and Economic Activity, 2000

Men

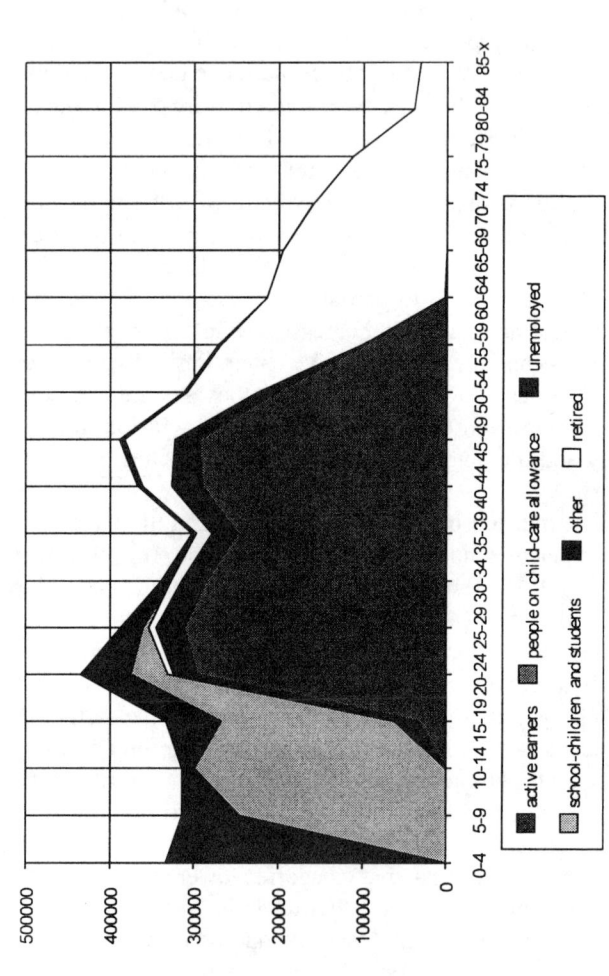

Demographic Transitions and the Change of Regimes in Hungary 195

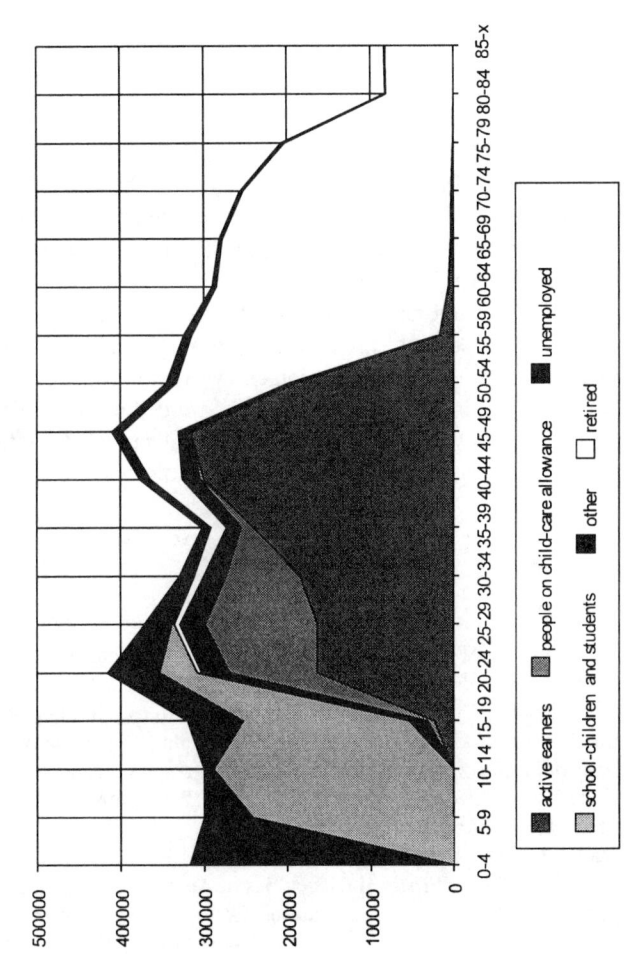

Estimated Number of the Population in Categories According to Sex, Age and Economic Activity, 2000

Women

There must be a mandate that besides short-range economic and social contracts, a new long-term social contract should be formulated. It should contain the program for self-realization for these generations, a new system of the income flow, and the overdue reform of income redistribution, first of all the pension system.

Summary

The present paper attempted to explain the demographic consequences of the change of political regimes in Hungary in the context of long-term demographic tendencies, and the conditions of the demographic transitions. In Hungary the fundamental demographical changes took place relatively early as compared to the economic development of the country, so the change of regimes did not result in so drastic demographic changes as in several other East Central European countries. At the same time, however, the prevailing demographic conditions may create obstacles in the way of the change of regimes, and urge for quick and thorough reforms in the welfare system. However, it seems that the realization of welfare reforms will be even harder than creating a new economic system in the forthcoming years.

Notes

1. Coale, A. J., "The Demographic Transition", in: *International Population Conference*, vol. 1 (Liége, 1973), 53–72.
2. Andorka, R., *Gyermekszám a fejlett országokban* (Number of Children in the Developed Countries), (Budapest: Gondolat, 1987); Dányi, D., ed., *Demográfiai átmenet Magyarországon* (Demographic Transition in Hungary), KSH NKI Történeti Demográfiai Füzetei (Papers on Historical Demography), no. 9 (Budapest: KSH NKI, 1991).
3. Valkovics, E., *A demográfiai átmenet elemzésének néhány gyakorlati nehézségéről* (Some Practical Difficulties of Analyzing the Demographic Transition), KSH NKI Kutatási Jelentései (Research Reports of the Demographic Research Institute of the Central Statistical Office) 5 (1982)(a)
4. Tekse, K., *Bevezetés a stabil népesség elméletébe* (Introduction into the Theory of Stationary Population), (Budapest: Statisztikai Kiadó Vállalat, 1975).
5. Van de Kaa, D., "Europe's Second Demographic Transition," *Population Bulletin* 42 (1984): 1–59.

6. Valkovics, E., "Az európai országok második demográfiai átmenetéről" (On the Second Demographic Transition of the European Countries), in: KSH NKI Hírlevél 5 (1992).
7. *Long-range Population Projections. Two Centuries of Population Growth 1950–2150* (New York: United Nations, 1992), 35.
8. Höhn, Ch., "Aktuelle Bevölkerungsfragen in Europa und in den anderen Indutrieländern", *Zeitschrift für Bevölkerungswissenschaft*, no. 3 (1992): 271–289.
9. Klinger, A., "Az öregedés demográfiai vonatkozásai" (Demographic Aspects of Aging), *Demográfia*, no. 1 (1983): 9–49.
10. Hablicsek, L., "A népesség korösszetételének alakulása, a kezelés problémái" (Changes in the Age Distribution of the Population and the Problems of Handling Them), *Demográfia*, nos. 1–2 (1991): 224–244.
11. Hablicsek, L., *A magyarországi demográfiai átmenet vizsgálata* (Analysis of the Demographic Transition in Hungary), in: KSH NKI Kutatási Jelentései 42, no. 2 (1992).
12. Hablicsek, L., *Magyarország népességének előreszámítása, 1993–2020. Az 1993. évi népességi és az 1992. évi népmozgalmi adatokon alapuló számítások eredményei* (Population Projections for the Period 1993–2020 in Hungary Based on Population Data from 1993 and Demographic Data from 1992), KSH NKI Demográfiai Tájékoztató Füzetek (Demographic Information Bulletin of the Demographic Research Institute of the Central Statistical Office) 14, no. 1 (1993).
13. Hablicsek, L., *Családok és háztartások előreszámítása, 1990–2010. Az 1992. évi népmozgalmi és az 1990. évi népszámlálási adatokon alapuló számítások eredményei* (Family and Household Projections, 1990–2010 Based on Demographic Data from 1992 and Census Result from 1990), KSH NKI Demográfiai Tájékoztató Füzetek 15, no. 1 (1994).
14. Hablicsek, L., *A gazdasági aktivitás előreszámítása, 1994–2010* (Projections of Economic Activity, 1994–2010), KSH NKI Demográfiai Tájékoztató Füzetek 16, no. 2 (1994).

László Cseh-Szombathy

The Role of Mental Elements in Demographic Phenomena

In the course of the twentieth century, the fundamental demographic processes like marriage, divorce, childbirth, and death in Hungary paralleled those in Western European countries. The only difference was that the changes began here later, around the middle of the century. At that time, the pace of certain demographic processes accelerated and Hungary took the lead in several respects.

The lag at the beginning of the century could be attributed to the general economic and social backwardness of Hungary, which was reflected also in the demographic processes. On the other hand, the changes after the 1950s had their roots in the Communist takeover and the ensuing political, social, and economic situation that exerted a decisive influence on the demographic factors. Although there were no such political changes in Western Europe, there were certain fields, like fertility, that followed the East European example with a certain time lag.

So the explanation for the Hungarian demographic processes goes back to a double root: to the economic and social backwardness of the country in the first half of the century and to the Communist dictatorship and its impact on society in the second. However, international comparison shows that these explanations are merely simplifications and leave out of consideration the mental changes in the minds of the people.

In one of his essays Philippe Aries[1] examined two centuries of changing fertility trends and the role of mental factors in them, and offered a theoretical model focused upon the changing ways of looking at children.

Aries puts the first significant change in the period between the end of the Middle Ages and the eighteenth century, when children became more significant both for the families and for society as a whole. This change was partly due to the economic expansion in those years which demanded expansion of the labor market. The children soon worked as "small adults" contributing to the prosperity of the family, and supplementing the adult labor force. The Middle Ages were characterized by population control,

Aries argues, even though its most brutal form was direct or indirect fetus murder. This was the situation that economic expansion changed for the better not necessarily as a result of rational consideration, but following from the changing public opinion.

The next change was brought about by the emotional appreciation of children, a trend beginning in the late eighteenth century. There were significant differences in the number of children among the various social strata, just as in the later periods. The change took place mostly among the middle strata that wished for a better life for their children and were willing to endure the increased costs of bringing them up. This concentration on children or, as Aries put it, this "kingdom of children" made birth control imperative, resulting in a steady fall of the number of births in an ever growing number of countries. Family planning did not characterize all layers of society to the same degree. For the poor there was little possibility of economic betterment, no matter how many children they had, and the living standards of the rich were obviously not influenced by the number of their children. It follows from this that the graph of birth control in the nineteenth century gives an U-shaped curve with the x-axis showing the social and economic position of the families, and the y-axis showing the number of children.

Aries maintains that family planning motivated by the outstanding role of children lasted till the 1940s, and the decreasing fertility of the total population resulted from an ever growing number of families trying to ensure social mobility by limiting themselves to one or two children. This was made easier by the appearance of the first generation of contraceptives.

Nor did the baby-boom of the 1940s and 1950s involve a fundamental change of attitudes, according to Aries. Children were considered "kings" in those decades, too, but the long-drawn economic boom convinced couples that they could provide their children with the education needed to reach an acceptable level of consumption and social support, even if not a rise in their social status.

Under these circumstances the women's movements expressed a dissenting opinion for the first time. The younger generations of women felt that giving birth several times would encroach upon their freedom and they would become sexual slaves of their husbands.

The last period, still in progress, began in the mid-1960s. The above-mentioned feminist organizations aiming at the liberation of women can be considered forerunners of the great change called the period of the second demographic transition. There is such a decrease in the number of births that it does not ensure the reproduction of the population any longer. The families either remain childless or have only one or two children.

Aries maintains that this new form of family planning cannot be attributed solely to the economic insecurity beginning with the 1960s in the western world. Families do not return to the model of having few children in order to be able to give them a reasonable education and thus make them competitive when they are on their own. The point here is that the priority of the children is over, they are no "kings" any more. Men and women in their procreative age do not necessarily envisage their future life with children. Should they still have some children, they play only a secondary role in their family.

This approach of Aries is new and harmonizes with his several decades of work in the field. His examples mainly characterize the French developments of the period and are taken mostly from French statistics of the subject. On the following pages I try to apply this model and test its applicability to the Hungarian situation, and I may even add to it.

Demographical changes in Hungary from the sixteenth to eighteenth centuries differed from the general trend and do not fall into Aries's first category. During the sixteenth and seventeenth centuries most of the country was under Ottoman rule or was subjected to a series of Turkish plundering expeditions. People were the most valuable items of booty for the Turks, and future slaves were dispatched to the core provinces of the empire. Even the population of the villages and towns already occupied was often carried off in spite of the fact that their taxes greatly contributed to the income of their Turkish masters. Children were future soldiers in the eyes of the Turks. Living under continuous threat made any kind of family planning impossible for Hungarian couples, and this insecurity influenced their fertility rate.

The second Turkish siege of Vienna in 1683, which ended in their fatal defeat, was the starting point of the campaigns to expel the Turks, which was accomplished in a peace treaty concluded in 1699. The campaigns lasted a decade and a half and caused further destruction. Even those villages lost their population that had survived the marauding campaigns of the previous centuries.

The size of the Hungarian population at the time of the expulsion of the Turks is a central question of Hungarian historical demography. As regards the applicability of Aries's idea, the important point is that a period begins when children become increasingly important from economic point of view, so high fertility comes to be viewed as a desideratum.

It took a century to repopulate the uninhabited regions of the country. This was partly an organized process in which foreigners from other countries were encouraged to settle in Hungary. On the other hand, those native Hungarians whose families had sought refuge in the peripheries, tended to come back to the southern and central parts of the country. Destroyed villages and towns revived with an impressive speed. A few decades later, several of these resuscitated settlements became starting points of a new inner migration, reviving the Hungarian character of the formerly occupied central parts of the country in most cases. This process was made possible only by the high fertility of the Hungarian population and a steady, biologically optimal number of children, which meant six to eight children per family.

This high rate of fertility lasted in Hungary up to the late nineteenth century. The first demographic transition must have arrived here late because mortality also remained high and families could be certain of a sufficient labor supply only if they counted on not losing half of their children before adolescence.

Certain villages or regions differed from the general trend. There was a radical reduction of births in those settlements as early as in the first third of the nineteenth century aiming to produce one child per couple. The existence of low fertility enclaves in a sea of high fertility can primarily be attributed to specific property relations at these places, but there was also an attitude similar to Aries' idea about ensuring a happier future for the children by reducing their number. Farms within large estates in Eastern Europe generally offered no opportunity for the landowning serfs to increase the size of their plots, so they could preserve their children's advantage over the landless ones by passing their land to one heir instead of dividing it among several. The single child of such families became once again "king" in the family. For these peasants, finding a similarly brotherless wife also meant a contribution to the family plot. These forerunners of birth control acquiesced in the economic factors behind their family planning, but they also considered it shameful and irresponsible to have more than one child.

Villages with one child in each family were exceptional in Hungary up to the late nineteenth century, and did not really influence the national average of forty per thousand and above up to the turn of the century, while the mortality rate was between thirty-five and thirty-eight per thousand. So the natural increase was rather low and started to increase only in the 1880s with the gradual decrease of mortality. This gave rise to the so-called first demographic transition in Hungary, too, beginning with the spectacular decrease of deaths. However, the rate of births also started to fall soon afterward.

The European waves of emigration to America in the early twentieth century affected Hungary as well and contributed to the spread of the idea and practice of family planning. World War I, breaking out in 1914, had, however, much more brutal results than a mere change of attitudes. The ensuing territorial mutilation of the country with its economic and political consequences, and the changing public sentiment contributed to an ever more extensive birth control in order to stop the lowering of living standards at least for the children.

The further decrease of the number of births in the 1920s and 1930s fully harmonizes with Aries's model in Western Europe. The only difference needing explanation is the lack of the baby-boom generation in Hungary after World War II, which was substituted by a forceful increase of births in the first half of the 1950s. After the failure of this attempt at increasing the number of the population another drastic fall in the number of births preceded the so-called second demographic transition.

The absence of the baby-boom generation in Hungary is probably due primarily to the political and economic events of the early 1950s. Soviet occupation and a dictatorial regime were established without any legitimate foundation. At the same time, the traditionally most productive rural population was systematically ruined. Women in their most productive age were forced to go to work, so the families were discouraged to have more than one or two children at a time when in the United States and in Western and Southern Europe families with three or four children were the average.

Aries's model does not touch upon the forceful intervention of the state and its effects on fertility, specifically the results of governmental campaigns against birth control between 1952 and 1955. It was an episode in the changing attitude and practice of power that was terminated partly by its failure to influence fertility to a desired degree, largely because of its negative impact on public opinion. After this period, birth rates tended to fall again. From 1959 onward, the total number of births indicated that the given birth rate did not ensure reproduction any longer, and the population was to decrease in absolute numbers within twenty to twenty-five years.

By this Hungary arrived at the threshold of the second demographic transition nearly ten years earlier than most Western European countries. The birth rate of 12.9 in 1962 was a negative record in Europe and gave rise to several explanations taking special local factors into consideration.

The idea of limiting birth control and that of extending more substantial state support to married couples were political considerations to increase and stabilize the willingness of the families to have more children.

The state intervention to improve demographics proved to be successful for four years in the 1970s, but after that the indices fell back to the level of the 1960s. In the meantime, low birth rates ceased to be a special Hungarian phenomenon, and the other European countries gradually sank below the Hungarian level of the 1960s. So it became obvious that the Hungarian trends could not be attributed solely to the local political and economic conditions, and the common European trends or changing attitudes were much more important. This was what Aries meant when he spoke of the end of the rule of children. Other authors speak of the spread of the post-modern world view and most of them agree that it was the change in attitudes and not the unfavourable turn in the living conditions that was of primary importance.

This attitude of the foreign demographers urged some of their Hungarian colleagues to examine fertility trends from the perspective of social values in the population. They did not wish to dismiss the experiences of the 1950s and 1960s, but wanted to avoid ethnocentric explanations for a phenomenon equally characterizing modern industrial societies directed by post-modern services and scientific knowledge.

In the last decade and a half, sociologists and demographers have extensively dealt with the above described change of attitudes. Their conclusions differ greatly, but they agree that there are no larger common values guiding families, and people are free to choose any of the possible alternatives according to their personal motivations.

Nor are generally applicable denominations and concepts describing these attitudes. Some give the term "post-modern" to families displaying the new types of attitudes, though the exact nature of this ideology in the changing demographical pattern is not quite clear yet. Most scholars speak of family values, and this definition can be accepted if one starts with the definition of the term "value" given by Clyde Kluckhohn, who maintains that the values are the explicit or implicit ideas of what is desirable that influence us in choosing the means and goals of our actions.[2] The use of the values in family sociology often comes into conflict with the widespread idea of Milton Rokeach emphasizing the abstract nature of values, which means that they are not linked to a specific object or place, express a global world view, and control action indirectly, by means of various attitudes.[3] These attitudes are, in turn, directed at a given object or situation, and are oriented both from a cognitive and from an emotional point of view. In works on family sociology the border between values and attitudes is sometimes rather blurred.

The basic concept of value analysis with Rokeach's method is that the most important values are shared by all and people differ only in ranking the general and common values. The uniqueness of a person's attitude lies in his scale of values. Some family sociologists think, however, in pairs of values excluding each other. Such are, for example, equality and inequality, systematic order and spontaneity, respect of authority and skepticism. The researchers following this method offer a choice for the interviewee. Talcott Parsons also maintains that values manifest themselves in our choices between the actual alternatives.[4] Aries's categories for establishing the value attributed to children in the various historical periods can also be included in these approaches.

To avoid misunderstanding, one must differentiate between alternatives as requirements or norms. We prefer to use Robin Williams' definition for "norms" as concrete specifications of preferred attitudes or behaviors, legitimated and linked together by the values behind them.[5] This means that a norm can express more than one value, and the same values can appear in several norms. Values are personal in nature, because internally motivated, while norms are conditioned socially, and have their roots in the institutionalized order and the roles they engender. In the case of norms, attached sanctions are also important, since following them does not come automatically, but depends on our expectations of sanctions in case of conforming to them or violating them. It follows from all this that the norms themselves represent a kind of order, at least the ones referring to a certain province of life like family attitudes.

Some contemporary family sociologists like the Dutch Bram Buunk, the German Kurt Lüscher and the French Louis Roussel[6] maintain that the norms regulating family life in the past two centuries also formed systems and corresponding ideologies. Aries's train of thoughts mentioned earlier also confirms this view. All of these authors admit that the satisfaction of an emotional claim in decisions within a family represents a differentiating factor in society and that the emotional functions tend to be pushed to the background.

One of the characteristic features of this attitude breaking away from feelings is the separation of sexual intercourse from emotional ties in the so-called open marriages. This new ideology of partnership openly declares that each partner has the right for a maximal sexual satisfaction disregarding the needs of the other party. It became a commonly accepted idea that relationships can only be transitional and temporary. Emotional attachment necessarily slackens with time and routine, and with inevitable conflicts. It is not worthwhile maintaining the stagnant relationship and keeping it up formally. The requirement of equality between the partners is being reinterpreted as well. It can be realized only if the family roles cease to be differentiated. The division of labor within a family should be spontaneous, occasional and instinctive. Experimentation within the family is permitted and encouraged to gain new experiences. Insisting on traditional methods results in disinterestedness. Family economy should also be guided by spontaneity. Planning which assumes permanent prospects is rarely justified, so satisfaction of personal needs should not be impeded. Family members should give enough freedom to one another in this respect, too. Loose relationship is also desirable between the family and the given society. The fewer norms that restrict the life of a family, the better. People are not responsible for the functioning, survival, and the reproduction of a society. The number of children in a family should be determined by its positive effects on the lives of the parents. The birth of a child creates an irreversible situation, and the post-modern concept of life prefers the maintenance of the possibility of a second chance as long as possible.

This new concept has been advertized widely both in the mass media and in special literature all over the world. The listeners and the readers may get the impression that the new norms have become dominant both in North-America and in Western Europe, and the norms of the so-called modern period arising from the first demographic revolution are cherished only by the older generation. These appearances are, however, contradicted by empirical surveys made in the last few decades concerning the changing norms and values in the families.

For example in the United States where the new norms were publicized most widely, surveys taken in the 1980s revealed that childlessness remained sporadic. Most of those interviewed condemned intentional childlessness. The image of childless couples is negative for the public.[7] Voluntary childlessness was mainly supported by feminist movements asserting that the positive approach to parenthood and the major role attached to motherhood were only means of maintaining the conventional sexual roles to the detriment of women.

The situation in Germany, mostly in former West-Germany, is characterized by the parallel functioning of various norms and their inconsistent application. The new ones have been accepted only by the minority, but the majority is highly tolerant towards attitudes deemed deviant only recently. The most widespread debates in Germany center around the role of these new norms in the low fertility rates. Research has found great discrepancy between the number of children the couples planned or considered ideal before their marriage and the actual number of children they had afterwards. This seems to prove that it is not the post-modern norms that lead to a decline in the number of births but the experiences during married life that modify the previous intentions. One of the most outstanding experts of the problem, Rosmarie Nave-Herz mentions the shock caused by the arrival of the first child. She says that the changes in the family following the birth of the first child create extra financial and physical burdens, a modified relationship between the new parents, and a resulting reaction to these changes that can prove to be discouraging.[8]

Sweden and Danmark were the first countries to adopt the new norms on a large scale. This, however, does not mean that the adult population follows them consistently throughout their lives. As Jan Trost, a pioneer of the research of concubinage writes, it is not unusual that partners legalize their relationship only after long years of living together with their common children taking part in the ceremony.[9]

The representatives of the new norms are most true to their principles in Holland. These flourished there a decade later than in Scandinavia, but they have been adopted by most of the young and the middle-aged generation. Nearly two thirds of the persons interviewed find nothing exceptional in voluntary childlessness. The rate of those remaining childless amounts to twenty per cent of those establishing new relationships.

The continuous research done by the Hungarian demographers presents an exact picture of the changes of family life and family attitudes in the last three decades, so the appearance and spread of the new norms can be observed and followed. Public opinion polls have repeatedly been conducted and reliable and rich information is available on demographic and family attitudes. This work was complemented by a survey done by Edit S. Molnár, Mrs. Tibor Pongrácz, and Ágnes Utasi in 1993 on the views, opinions and lifestyles of men and women in their productive age.[10] The survey also contained questions about family norms and was aimed at verifying the hypothesis that post-modern views had become wide-spread in Hungary by the last decade of the twentieth century within the population of productive age, thus giving rise to low fertility rates.

The interviewed young and middle-aged people were asked to choose between eight pairs of alternatives referring to problems in the family. The first set of statements expressed the conventional norms of the first half of the twentieth century, while their alternatives expressed those of the past few decades. All of these statements reflected the crucial points of the change of attitudes. The first pair confronted thoughtfulness or consideration with spontaneous action and questioned the significance of rational thinking. The second pair offered a choice between time-honoured solutions and experimenting. The third was about self-control and the free pursuit of personal desires. The next pair confronted self-sacrifice and the representation of one's own interests, then came the observance and the refusal of authority. The sixth pair of statements spoke of observing public opinion and the free display of one's own inclination. The seventh pair confronted the demand for sincerity and openness with guardedness or reserve. Finally, the eighth question wished to know if one preferred consumers' society or one where material wealth was no longer an aim in itself.

Most interviewees preferred the traditional norms in seven cases, the most popular being self-sacrifice. The rate of men and women did not differ in this respect, ninety per cent of both sexes considered it their duty to make sacrifices for their partners. The least popular was the traditional idea of following time-honoured norms instead of experimenting (less than fifty per cent). The most popular "post-modern" idea voted against openness in marriage. Forty per cent maintained that it was unsafe to be totally open with the partner and the partners had the right to have secrets of their own.

The above statements proved suitable for indicating the differences of opinion in the public. They reveal that the majority was still loyal to the traditional norms in Hungary in 1993, but there was a significant minority preferring dissenting views. It is interesting to know how these two types of opinions correlate. Calculations revealed that there was not a strong correlation between the two. Therefore one concludes that there is possibly no definite post-modern world view behind the answers of the minority to account for the differences of opinion. So it cannot be proved that the spread of post-modern norms should contribute to the great number of dissenting views and attitudes concerning family life, and to low fertility rates. Still, post-modern norms persist among the young and middle-aged population. The norms do not unite into a general ideology, but they might in the near future. Should the state prefer to strenghten the traditional forms and functions of the family, it should be prepared to a debate over world views, besides establishing indipensable economic, employment, and educational programs for the future. This debate cannot be won authoritatively, but it may suggest that traditional forms are capable of modification in our world.

Notes

1. Philippe Aries, "Deux motivation successive du declin de la fécondité en occident", in: *Seminar on Determinants of Fertility Trends* (Hamburg, 1980).
2. Clyde Kluckhohn, "Values and Value Orientations in the Theory of Action", in: T. Parsons, A. Shils, eds., *Toward a General Theory of Action* (Cambridge: Harvard University Press, 1951).
3. Milton Rokeach, *The Nature of Human Values* (New York: The Free Press, 1973)
4. Talcott Parsons, Edward Shiles, "Values and Social Systems", in: Jeffrey C. Alexander, Steven Seidman, eds., *Culture and Society. Contemporary Debates.*
5. Robin M. Williams, "Change and Stability in Values and Value Systems, a Sociological Perspective", in: Roheach, ed., *Understanding Human Values, Individual and Social* (New York: The Free Press, 1979).
6. Bram Buunk, "Alternative Lifestyles from an International Perspective", in: Eleanor T. Macklin, Roger H. Rubin, *Contemporary Families and Alternative Lifestyles* (Beverly Hills: Sage Publications, 1983).
7. Jean Veevers, "Voluntary Childlessness: A Critical Assessment of the Research", in: Eleanor T. Macklin, Roger H. Rubin, op. cit.
8. Rosmarie Nave-Herz, "Kontinuität und Wandel in der Bedeutung, in dem Struktur und Habilität von Ehe und Familie in der Bundesrepublik Deutschland", in: Rosmarie Nave-Herz, *Wandel und Kontinuität der Familie* (Enke Verlag, 1988).
9. Jan Trost, *Unmarried Cohabitation* (Verteras International Library, 1979)
10. László Cseh-Szombathy, Edit S. Molnár, Mrs. Tibor Pongrácz, Ágnes Utasi, "Családi értékek, családi normák" (Family Values, Family Norms), *Magyar Szemle*, no. 9 (1994): 918–932.

Marietta Pongrácz and Edit S. Molnár

Adolescent Fertility

Pregnancy among women under eighteen is very frequent in Hungary even by international standards. In 1983 a survey was conducted to establish the causes and the consequences of adolescent fertility, the family background, and the social and demographic characteristics of the teenagers involved. In 1993 the same young women were interviewed about their lives and their children. The present study contains the most important conclusions of this survey.

Adolescent pregnancy, childbirth, and abortion have recently become the center of interest in various developed industrial countries. The increased interest of demographers, sociologists and physicians can be traced to the fact that the frequency of adolescent pregnancies is very different in countries on similar social and economic levels and among teenagers coming from similar social and economic backgrounds. The outcome of the pregnancies is also very different, so the rate of births and abortions may be very different in the individual countries.

An international comparative survey published in 1983[1] establishes that out of twenty-nine countries examined only Romania and the black population of the United States exceeded Hungary in adolescent fertility. In all other countries from New Zealand to the United States, teenage fertility was significantly lower.

The Hungarian situation was further aggravated by the fact that while in Western Europe and in the United States women under twenty are considered adolescents, in Hungary the age limit is eighteen from a demographic point of view. Statistically, women of 18 and 19 are included as adults. In other words, it is the Hungarian age-group 15–17 that should be compared with the age-group 15–19 in most European countries.

The aforementioned international survey called attention also to the fact that Hungary is closely followed by the United States in adolescent fertility. It is remarkable that two countries so different in their size, economic development, and history should be so similar from this respect, even though this problem is far from being merely demographic. The high frequency of teenage pregnancy has placed the problem into lime-light in the United States, and many demographic, sociological, and psychological studies try to find the causes and reveal the consequences. American experts have found that teenage pregnancy is more frequent in countries where

- there are many married women in the age group 15–19
- there are many abortions
- the rate of the agrarian population is high
- the factor of religion influences the behaviour of the population
- the state grants substantial benefits and subsidies for mothers
- the rate of economically active women is high
- there is a supportive pro-natal policy at the state level
- consumption of alcohol is high.

Most of the above factors are valid for the Hungarian population as well. The rate of married women in the age-group 15–19 is very high, many times higher than in the neighboring countries and in Western Europe. The rate of abortions is also very high, as is the percentage of women in the agricultural population.

The role of religion in the life of the Hungarian population is difficult to define. Prior to World War II, Hungary was a deeply religious, and an overwhelmingly Catholic country. During the past forty years of Communist rule, the significance and social impact of the Catholic Church greatly diminished, but the number of religious people remained significant. Sociological research has revealed that the rate of people calling themselves religious has been, and still is, around 60 per cent. This does not mean the daily exercise of one's religion, but rather the acceptance of religious values and adhering to a more conservative view of life, especially in sexual matters. In this sense, both the Hungarian and the American society show an equally ambiguous attitude concerning sexual life and the sexual behaviour of the young.

While the mass media, films and literature are fairly open about adolescent sexuality, the parents and families involved either brush the problem aside or advocate conservative views that are incongruous with current mores. This ambiguous, and often rather conservative attitude may also contribute to the great number of adolescent pregnancies in Hungary. (The basically conservative values of the Hungarian society have been supported by sociological research.)

The correlation coefficients of the international survey suggest that an actively pro-natalist state policy supporting childbirth in a country and the widespread support of mothers also contribute to high teenage fertility, just like the high rate of economically active women. In Hungary, 81 per cent of mature women work outside their homes, which is higher than the international norm.

Countries with a high comsumption of alcohol and adequate statistics to measure consumption (i.e., the number of those dying of cirrhosis) also show a high rate of adolescent fertility. Unfortunately, Hungary occupies a prominent position in Europe in this respect, since the rate of death due to cirrhosis is very high, and continues to rise.

Besides analysing the factors influencing adolescent fertility from a mathematical and statistical point of view, the above mentioned American survey dealt also with numerically undefinable factors, such as the teenagers' knowledge about contraceptives and their cultural level. The experts established that high adolescent fertility in the United States is partly due to the accidental nature of teaching the relevant information, to the lack of systematic family planning, and consequently to the low rate of young people using contraceptives. As regards birth control, Hungary faces similar problems. Schools have been dealing with problems of contraception in the framework of a subject called "family life" since 1974, but relevant information does not influence the teenagers to the desired degree. Although statistical data are not available concerning the sexual behaviour of the young or their use of contraceptives, it is highly probable that the Hungarian youths are uninformed in this respect.

To sum up, the results of the international survey can be taken as relevant also to the Hungarian demographic circumstances, and the factors contributing to adolescent fertility in the developed countries are valid for the Hungarian society, as well.

Changes of Adolescent Fertility in Hungary

Table 1
Live Births and Abortions of Adolescents

Year	For thousand women in the individual age-groups								
	Live births					Abortions			
	-14	15	16	17	15-49	-14	15	16	17	15-49
1975	1.9	9.3	24.6	56.0	72.8	1.8	6.0	15.4	25.4	36.1
1978	2.8	10.6	29.8	62.9	64.1	2.1	6.7	17.7	28.5	31.9
1983	2.6	9.3	24.2	47.4	49.8	2.3	7.3	17.4	29.0	30.8
1985	2.8	9.7	24.1	47.2	75.5	2.0	7.3	17.0	30.1	32.0
1990	1.4	6.5	17.7	34.0	49.4	3.3	10.5	20.4	33.3	36.0
1993	1.3	6.4	15.3	26.8	45.3	4.0	11.2	21.4	33.0	29.0

The data in Table 1 show that the rate of adolescent pregnancy was very high in the seventies and the eighties and it is high even today. It was identical in the early eighties and in the early nineties, namely, thirty in every thousand young women under eighteen got pregnant both in 1983 and in 1993. There was, however, a difference in the outcome of the pregnancies with an increasing number of pregnancies ending in induced abortion in 1993. In 1983 eighteen pregnant teenagers gave birth to their children and twelve had their pregnancy interrupted, while in 1993 thirteen teenage mothers gave birth, and seventeen resorted to abortion.

Table 2
Number of Teenage Pregnancies According to Outcome

Age	1983		1993	
	Live Births	Abortions	Live Births	Abortions
-14	189	172	199	303
15	667	515	522	884
16	1584	1160	1307	1783
17	2957	1836	2399	2807
Σ	5397	3683	4427	5777

Chart 1
Distribution of Teenage Pregnancies According to Outcome in per cent, 1983 and 1993

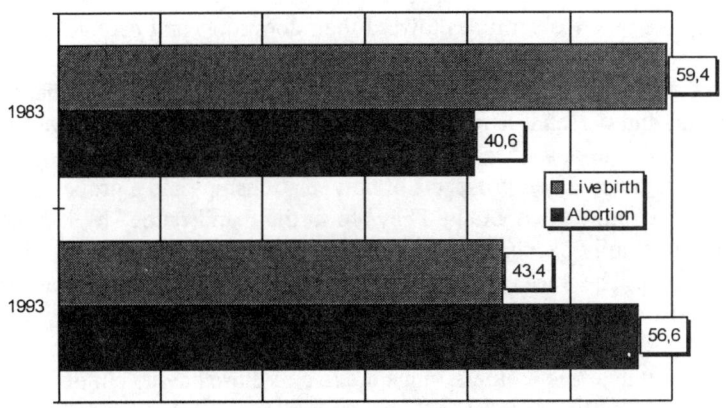

The question is what social and individual consequences live births and abortions in adolescence have and how they should be interpreted.

Experts have widely differing opinion in the subject. They consider childbirth in such a young age very negative socially and unfavourable as a social and demographic phenomenon. Expert opinions concerning procured abortion are less unanimous. Some think that the risks of carefully executed and psychologically well prepared abortions in due time do no damage from a health or psychological perspective. Others maintain that the biological and psychological consequences are far from being negligible.

Survey One[2]

The high and not abating frequency of adolescent pregnancy has made it important to conduct surveys to find out the causes, factors, and consequences. Our survey took place in 1983 and all women under eighteen experiencing childbirth or an abortion in the first half of the same year were interviewed. Out of the sample of 3448 young women 1135 were unmarried undergoing abortion, 1393 were married at the time of the childbirth, and 950 gave birth out of wedlock.

In the first group of women our questions referred to demographic factors such as family background, adolescent experiences socialization, information about their partners, sexual relationships, sexual conduct and contraceptive practice, attitudes toward their abortions, and parental reactions.

The situation of teeenage mothers keeping their children is burdened with social and personal hardships. Most young mothers do not have the proper knowledge and education to look after their children properly or finish their studies. Their prospects of new relationships and a proper family life in the future are dubious. They are further handicapped by lacking a supportive family background, a tolerant neighborhood and a supportive social system. The variegated social consequences of childbirth make a more detailed survey necessary as regards the social, demographic, and sanitary background. We have found it advisable to distinguish between married and unmarried mothers, since the rate of those having children out of wedlock is nearly one third of the total number of mothers under eighteen.

However, nearly three quarters of all under-age married mothers got pregnant before they got married. Marriage means social security and legality for them, while the situation of unmarried mothers, who sometimes do not even know who the father was or have long broken their relationship with him, is less acceptable by society.

Besides the basic demographic factors we examined in both groups the family status of the mothers, their upbringing, the factors influencing their socialization, and certain circumstances influencing their life standards such as housing and income, information about the baby's fathers, and the duration and circumstances of the relationship. In the case of unmarried mothers we inquired about the perspectives of the relationship with the father, and the allocation and support of the child. In both groups, we tried to find out the level of the couples' knowledge of contraception, their relevant practice, their reasons for keeping the baby, the reception of their decision in their larger family, and the opinion of their parents. The teenage mothers' plans for their future family life, education and work were also represented in the questionnaire. Last, but not least, we recorded the course of pregnancy and delivery, the possible complications, and some important sanitary and demographic characteristics of the babies.

Our results are the following. We found that mothers under eighteen do not constitute a homogeneous group. Early sexual life and teenage pregnancy are common, but the circumstances and motivations of their impregnation, their subsequent behaviour, and their decision about keeping or aborting the baby are significantly different. All these differences of attitude go back to differences in their social background and demographic factors. So the research results bear out our preliminary hypothesis that pregnant teenagers can be grouped into three categories on the basis of the outcome of their pregnancy, and their family status. The only modification in the hypothesis is that the even group of mothers having children out of wedlock can be divided into two sub-groups, namely, to the homogeneous group of those living with their partners and to the heterogeneous group of single girls, much less definable socially and from demographic point of view.

Adolescents having their pregnancy interrupted do not constitute a marginal group as regards their social and family background. They fit into society like any others. The level of their education is the same as that of others in their age-group, most of them learn in secondary schools. The rate of Gipsy population is low among them. The distribution of their dwelling places according to the type of settlement is also very similar to that of the average for the age-group 14–17. Parents of adolescents with interrupted pregnancies are also typical. Their fathers' education is close to the average, and they usually are skilled manual workers. Their mothers mostly work outside their homes, mainly in administrative jobs. The rate of these working mothers equals the national average. The fertility of the mothers of adolescent mothers does not differ significantly from the national average if the age at their first pregnancy and the number of their children are taken into consideration.

The higher cultural level and more conscious attitude of pregnant teenagers are shown by the fact that most of them use contraceptives, however unsuccessfully. The pregnancy is detected early or just in time and a medical exam is not put off. Thus 90 per cent of the abortions take place before the critical twelfth week. Their decision to have their pregnancy interrupted indicates their wish to continue their studies and lead a fuller life, though the attitude of their parents resenting a marriage of convenience also plays a part.

However, this picture of teenagers with induced abortion is valid only in a general way. It gives a valid image of most adolescent girls in question, but there are obviously teenage girls or groups of girls who do not fit the pattern, which is well illustrated by the statistical analysis of the relevant data. These girls are mostly very young, still in primary school. When they become pregnant, they leave school and do not work, either, and are often of Gipsy origin. Their sexual relationships are unstable and thoughtless, they often pursue random acquaintanceships. They usually do not know even the basic things about their partners and do not know who the father of their baby is. Many of them notice their pregnancies late, visit the doctor late, so their abortion may take place only after the thirteenth week or even later than that, sometimes only after the seventeenth. Although the rate of this group is not high, it is worth mentioning it, since its social and demographical background differs significantly from the average circumstances of teenagers with induced abortions that coincides with the national average.

Married adolescent mothers cannot be considered deviant, either as regards their social and demographic characteristics and their attitude to fertility. Although their education, family background, and the social status and occupational distribution of their parents falls short of the similar features of single teenage women undergoing abortion, this does not mean that married adolescent mothers are in a marginal position in society. However, since they come from social strata below the national average, these positions are likely to determine their future, too. Their level of education remains below the average, they usually leave school and go to work very early, and have unskilled manual jobs in want of a proper education. They come from a layer of society where early marriage and childbirth are natural and accepted or even encouraged by the families. These observations are supported by the responses of these mothers concerning their evaluation of their situation and their plans for the future.

Housing is the most important problem of these teenage mothers and also their financial difficulties. Early pregnancy and childbirth themselves do not represent a problem for them at all. Their ideas about the number of ecpected children match those of young married women in general. Twenty per cent of the interviewed mothers was satisfied with one child, two thirds wanted two children and eleven per cent wanted three. These rates are roughly identical with the results of the longitudinal fertility research. The question is whether these plans for the future are realistic and how they will come true at all. Our investigation of families with three or more children and the results of other experts lead us to the conclusion that women who give birth in their adolescence for the first time, have eventually more children than the national average. So these women usually exceed their original expectations.

While the positive pole of pregnant adolescence seems to be the group of those who have their pregnancies terminated, the other extreme is undoubtedly the group of adolescent mothers living in concubinage with their partners, who are at a disadvantage from the beginning. This group is highly homogeneous as regards both its members' social and family background, and their demographic characteristics. Their level of education is generally very low, they are usually two schoolyears behind the average of their generation. Many of them leave school as early as their elementary school years (in the first four years of primary school), do not work, either, and live as dependents. Their parents' demographic characteristics (their level of education and their occupational distribution) reveal a still more disadvantaged background. The rate of illiterates or people never having attended school is surprisingly high among these parents. Most fathers are unskilled workers, and the majority of the mothers are housewives. Most of the young mothers in this group are Gipsies by origin, but the non-Gipsies also show similar characteristics. The low cultural level of their families, the non-stimulating but example-setting environment determine their sexual attitudes and their attitude to childbirth. This group is the first to start sexual life, refuses to use contraceptives in most cases, so they are the youngest when their first child is born. They usually want to have babies, and since living in concubinage is an accepted way of life in this layer of society, their decision is supported by their family, too.

The plans of these mothers indicate that they wish to live in the same framework in the future, too, and only a few of them wish to legalize their partnership by marriage. When asked about the planned date of the marriage they were so uncertain about it that their sincerity seemed highly questionable. Their future fertility is similarly uncertain. They wish to have more babies than the young married mothers in general, and their actual fertility is sure to be even higher than planned. On the basis of our research on mothers with three or more children it is established that a significant portion of these mothers also had their first children in their adolescence and lived with their partner. It is, therefore, very likely that the group of adolescent mothers having children out of wedlock and living in concubinage constitute a layer that is going to be a very disadvantaged sector of the population with respect to educational opportunities and employment.

While these three groups of teenage mothers are homogeneous from the point of view of their social and demographic characteristics, and their adolescent attitudes, the fourth group of single mothers having children out of wedlock is highly heterogeneous. It equally comprises semi-illiterate gipsy girls and girls coming from well-to-do families living under better conditions than the national average who are forced to leave school because of their pregnancy. There are differences in their decisions for childbirth, since there are some who consciously undertake it as part of their life strategy. However, most of them chose to have the babies only because it was too late for terminating the pregnancy, which is a feature characterizing only this group of adolescent pregnant mothers. More than one fifth of them did not think of the possibility of pregnancy even between the thirteenth and sixteenth week of it. About seventy per cent saw a doctor after the twelfth week, and there were thirty-five mothers who first met a doctor only in the delivery-room. These girls seemed to deliberately ignore their pregnancies, they dared not face the facts and kept putting off going to the doctor through indifference or expecting some miracle to happen.

It is also characteristic of this group that the mothers' relationship with the father generally comes to an end before the baby arrives, and the fathers refuse to accept their paternity. All these factors create a very difficult situation both for mothers and children. Although relatively few of these teenage mothers renounce their children after birth, their difficult circumstances are likely to force many of them to offer them for adoption or put them in state custody. Adolescent pregnancy and childbirth seem to have the most serious consequences for these single mothers and the story of each is a personal tragedy.

Survey Two[3]

Experts often call attention to the long-term negative effects of adolescent childbirth influencing the life of the mother, her demographic behaviour, and her social status throughout her life. Longitudinal surveys serve to explore these long-term effects by interviewing the same persons repeatedly. In our case, it was the mothers of 1983 who were asked once more in the framework of a follow-up. Since we have not read about such a survey in demographic literature, we consider ours a pioneering study.

Ten years seem to be enough to establish a life course for these young women, to see how their family status had changed, how many children they had, how they were able to bring them up, and how they were integrated into society. However, omitted were those who had their pregnancies terminated in 1983, since it is considered to be morally unacceptable to remind them of their abortion and disturb their family life. These young women, either married or unmarried, who had a baby in 1983 were interviewed. It was not easy to find the addresses, but finally 66 per cent, that is, 1546 persons of the 1983 sample were found and interviewed. Their distribution is shown in the following table (in per cent):

	1983	1993 according to family status in 1983	Found
married with father of first child	59.4	61.7	68.6
living in concubinage with father of first child	28.8	29.2	66.9
not living together with father of first child	11.8	7.2	40.2
unascertainable	–	1.9	–
Total (per cent)	100.0	100.0	66.0
Total number	2,543	1546	

One may conclude from this that it was impossible to find just the most desolate, the most disadvantaged who got into marginal positions, and are, therefore, underrepresented. Although this may be valid, especially with those who had their children without a husband or a partner, it is contradicted by the fact that the rate of gipsy mothers, who were in the worst position in 1983, was practically unchanged in 1993. Still, greater attention was paid to this problem and subgroups were made according to family status in 1983. So a reliable picture of the changes in the previous decade is hoped to be offered.

The lives of these mothers have been followed up from several respects. While examining the mothers' family status at the time of the childbirth (1983), the question of legalizing their relationship with the father, and the duration of their relationship with the father, six basic types were set up:

Type 1: the mother was married at the time of the delivery of the first child in 1983 and was still married to the father of the baby in 1993 (50 per cent of the sample).

Type 2: the mother lived together with the father of the first baby in concubinage in 1983 and later they got married and were still married in 1993 (17 per cent).

Type 3: the mother lived together with the father of the first baby in concubinage in 1983 and the same relationship was still in existence in 1993 (4 per cent).

Type 4: the mother was married in 1993, but not to the father of her first baby (14 per cent).

Type 5: the mother was married in 1983, but in 1993 she was either divorced or widowed.

Type 6: the mother was single in 1983, never lived together with the father of the baby, and was still single in 1993 (5 per cent).

These types also give information about the upbringing of the children. Children in the first three types live in full families with their own parents. Those in Type 4 live in a full family, but mostly with only one of their parents and usually with half-brothers or half-sisters, while those belonging to Types 5 and 6 live with their mothers only. The above figures suggest that the children of mothers having their first children in adolescence are usually brought up in normal families with two parents, just like the national average.

So it can be established that the group of teenage mothers became largely homogeneous as regards family status in the ten years after their first pregnancy. Hardly more than 9 per cent of them remained single, about 81 per cent was married, 9 per cent divorced, and not quite 1 per cent was widowed at the end of the decade. One of the major motives of getting married was naturally the wish to legalize the relationship with the father. Even one third of the formerly single girls managed to marry the father.

Another important motive of getting married after the birth of the baby was to live a settled life, even though the mother had been left by the father of the first baby. Many of the formerly single mothers got married during the ten years in question.

The wish to get married seems to be general also in the case of gipsy mothers, and their marriages are not less stable than those of the average of their generation. Lawful marriage seems to give prestige to the woman also among the Gipsy population, partly because it offers financial security, and partly because it helps develop a favourable image of the family in the neighborhood. Other features like a good job or high income do not seem to be enough to improve their status and prestige in society.

It must be remembered, however, that this group of adolescent mothers is more likely to live in concubinage than the national average. Although it is more frequent among Gipsy mothers, it is also common among others, as is proved by the existence of Type 3.

It can also be established that the mothers found in 1993 did not differ significantly from the average as regards their attitude to marriage and divorce. It is only their early marriage and its longer duration that makes them different from those not having children so early. Their behaviour later is practically identical with that of their age group or social layer in general.

The social distribution of adolescent mothers shows that they lag behind their generation owing to their low education, the lack of marketable skills, the great number of children, and the higher rate of non-employed housewives among them. Even their moderate plans for finishing their studies and taking part in some kind of vocational training in 1983 were fulfilled only in unusual cases. Moreover, the husbands or partners were on a lower educational level than their respective male age-groups. At the time of our investigation, the rate of the unemployed was very high among them. So the social distribution of the new families of adolescent mothers was essentially the same as that of the families of their childhood. Breaking out of this environment would have been possible only through education, or a more settled partner with a higher income. The low educational ambitions of the mothers significantly contributed to their disadvantaged situation. Early childbirth and its absorbing consequences are not the only excuses for lacking plans for further education. They had not felt like learning and going to school, anyway.

The changes in the family status of these women and their repeating patterns of their own childhood suggest that adolescent fertility did not create more unfavourable situation for them. On the contrary, starting a family was according the norms and traditions of their social group.

The situation is the same if we take the number of the children of these young mothers into consideration. The majority came from families with several children, and their fertility in the ten years that passed since the first survey largely exceeded the average of their generation. Forty-one per cent on them had three or more children, while the national average in the case of women of 20 to 24 was 3.5 per cent, and in the case of the age-group 25–29 10.7 per cent. The average number of children was exceptionally high among those living in concubinage, among those living in concubinage at the time of the birth of the first child, and among Gipsy women.

Already in 1983 it was pointed out that owing to the biological immaturity of young mothers, more complications occurred, and the rate of immature, low-weight babies was higher than usual. Interestingly, these unfavourable biological parameters coincided with social ones, and they occurred more often among single adolescent mothers than among married ones. Our research confirmed these earlier reports of 1983. Premature delivery starting before the thirty-seventh week of the pregnancy was also more frequent than the national average, and is still higher among certain social groups, like single mothers, mothers on a very low level of education, and Gipsy mothers. All these indicate lasting unfavourable social conditions. Problematic birth (i.e., premature birth, and birth with a low weight) has a disadvantageous effect on the children's school achievement. The rate of handicapped (primarily mentally handicapped) children is high in the most disadvantaged group.

The children of adolescent mothers are usually brought up under normal conditions. In spite of the fact that nearly 25 per cent of them were born out of wedlock, most of them live with both of their parents or at least with one of them, i.e, in a family. Contrary to what was anticipated, the rate of those passing into state custody or to adoptive parents is not higher than the average, and the number of paternity suits is also very low. It is, however, certain that the first children of adolescent mothers were more often brought up without their father or in different homes, which was partly caused by the fact that nearly 28 per cent of the mothers broke their relationship with the father during the ten years in question.

Most children spent the first three years of their lives at home with their mothers. The rate of children attending nurseries was comparable to the national average, but was lower than what would be desired. Many mothers with several children and a low level of education, and those who do not work outside their homes think that they are at home anyway, and do not send their children to the nursery-school. Their lack of early formal education manifests itself later in their poorer achievement at school.

The social background of the parents manifests itself also in the greater number of failures at school and the more frequent school-year repetitions. At the time of the second survey, even the oldest children were still in the lower grades. Their failure to cope right at the beginning is rather alarming, since it repeats the maternal pattern, namely, that they will also become over-age and leave school early.

Having drawn the lesson of their own life, former adolescent mothers maintain without exception that it is very important to give the children the necessary information about the facts of life and the ways of contraception, and they wish to do it in due course. Most of them consider it to be the task of the parents, particularly the mother. It is worth noting that their own contraceptive behaviour in this respect has become fully identical with the female population in general in terms of regularity and method.

As regards the evaluation of their early pregnancies, these women do not consider their teenage motherhood exceptional or exceptionally detrimental to their later years. Taking care of their first babies did not seem to have been too difficult, except for sick children, and neither did the household chores. In the beginning many of them lived with their parents, most of whom helped the young mothers, however angry they had been because of their unexpected pregnancy.

Ten years before, the teenage mothers were most worried about their finances and the lack of independent housing. At that time we thought they were too inexperienced to gauge the consequences of their early childbirths. However, the second survey proved that they had been quite realistic. They still find the financial and housing problems to have been the greatest, though their "lost adolescence", the lack of recreation, and overwhelming childcare also were mentioned.

Family status at the time of the birth of the first child had long-range and diverse effects on the lives of the young mothers.

Those who lived with their husbands ten years before and still do were the most successful in leading a balanced and consolidated life. These couples were on a somewhat higher level of education, had a better family background, and only few of them were of Gipsy origin. Their housing conditions were good, and the children were brought up under normal circumstances.

Similar was the situation of those who sustained their relationship with the father. They lived in concubinage with the father at the time of childbirth, but later married them. The differences arise from the much more disadvantaged social background of the pregnant adolescents (e.g., lack of education, Gipsy origin, neglect of work, etc.). These factors do not cease with the marriage and bad housing conditions, larger number of children, and their poorer upbringing manifest them. The above applies even more to those who still live in concubinage with the father. These mothers are still less educated, many of them never work in their lifetime, have even more children, they often lead loose lives, and their children are often given to state custody.

In the 1993 sample a relatively small group was constituted by those who had been single when they had their first babies and still were. (The representation of this group might be higher if all of them had been contacted at the time of the second survey.) Many of them migrated to cities, mostly to Budapest. Some lived in concubinage several times during the ten years, but many of them still live with their parents. Although they try to bring their children up under decent circumstances, to which most of them are entitled to get an allowance also from the father, many of them do not live with their children any longer.

Mathematical and statistical analyses have shown how much the objective factors affecting the life of these mothers contributed to the great diversity in their ways of living. There are typical careers and also particular ones not fitting the pattern. It is even more difficult to establish a pattern when the young mother remarried, divorced or widowed. In these cases problems like housing difficulties or the allocation of children are naturally more numerous, so the mothers often respond on the questionnaire that their "problems have not been settled yet."

Serious personal problems may aggravate the situation still further. Some parents have still not become appeased, so the young mothers cannot rely on their help. Some chose a bad partner or were forced to get married. Some husbands were alcoholic and aggressive. There were also mothers who could not find a suitable partner just because of their having a child. These circumstances are directly or indirectly in connection with the early childbirth, but not only with that. Alcoholism, aggressivity, or vagrancy as problem-solving activities are more characteristic of this social layer than of the rest of the population.

It is a common feature of the former adolescent mothers that they are more dissatisfied with their financial circumstances and their lives than their age-group in general. If they could start all over again, they would have their first children at the age of 18 to 20, and those having three or more children would have fewer. The ideal picture of starting a family at a mature age and two or maximum three children is still in their minds. Most of them do not realize even after ten years that the lack of education, work-skills, and sexual culture would limit their chances in society, anyway.

Discussion

Seeing the results of the two surveys the question arises what society could do at influencing adolescent fertility, and if it is possible or necessary to intervene at all, and if the response is yes what ways and directions the interventions should be. The other important question is what social support could be given to the families with biologically and psychologically immature mothers, and if the disadvantages of children growing up in them can be diminished.

In most foreign countries authorities and experts cooperate in suppressing adolescent fertility. Some foreign experts maintain that it is not the age of the teenagers at the time of their first sexual experiences that matters but their attitude to contraception. In other words, the risk is not in sexual life begun too early, but in thoughtless behaviour.

The other key issue is the unwanted pregnancy. Foreign experts classify induced abortions and childbirths out of wedlock as unwanted pregnancies, while childbirths in a family are considered as desired, although not without reservations. Their primary aim is to suppress the number of childbirths out of wedlock, but in the case of abortions their opinions greatly differ.

Some consider abortions as harmful, therefore necessary to be suppressed, while others deem the biological and psychological consequences negligible and consider social intervention unnecessary.

The problem areas of teenage pregnancy, intervention and suppression need to be approached somewhat differently in Hungary. The first survey revealed that pregnant teenagers under eighteen are very heterogeneous both socially and demographically, and as regards their motives in becoming pregnant. The demographic approach to the problem also has to be very differentiated both as regards attitudes and methods. The situation and behavior of a teenage mother giving birth to a child in a steady or married relationship should be considered differently from those who give birth as a result of accidental circumstances or their thoughtlessness. The problem of desired and unwanted pregnancies is equally different from the international comparative examinations. In Hungary, most births out of wedlock were results of desired pregnancies and the partners live together in concubinage, so the categories foreign experts use are not quite applicable in this country. Pregnancies ending in procured abortion and those of single adolescents can be considered as unwanted in Hungary, too. However, pregnancies of married women and those living in concubinage should be considered desired ones, even if not all of them are really planned.

Making the right distinction between desired and unwanted pregnancies is important not only from theoretical point of view, but also with respect to the practical implications. While in the case of unwanted pregnancies preventive measures are important both for the individual and society, thus social efforts are more likely to be accepted by young people, the case of so-called desired pregnancies is different. Although most experts consider adolescent fertility harmful on the basis of the immaturity of teenage mothers both biologically and psychologically, and their pregnancy and childbirth disadvantageous socially, married adolescent mothers and adolescent mothers living with their partners cannot be denied the right to choose to be mothers if it acceeds with the norms and traditions of their families and neighborhood. It is questionable, therefore, if it is possible or even permissible to intervene morally or legally into the private spheres of such a unique subpopulation. The result of such an intervention might be counterproductive, anyway. Tolerance should be the most important feature of the attitude towards married adolescent mothers and those living in concubinage. Society should understand that there are social groups that consider it natural to get married early, live with partners early and have children in adolescence.

Notes

1. Westoff, Ch., Calot, G., and Foster, A., "Teenage Fertility in Developed Nations, 1971–1980," *International Family Planning Persepectives* 9, no.2 (June 1983): 45–49.
2. Pongrácz, M., S. Molnár, E., *Serdülőkori terhességek társadalmi-demográfiai vonatkozásai* (Social and Demographic Aspects of Adolescent Fertility), Népességtudományi Kutató Intézet Kutatási Jelentései 32 (Research Reports of Institute for Demographic Research), 217.
3. Pongrácz, M., S. Molnár, E., *Serdülőkorban szült anyák társadalmi-demográfiai jellemzőinek longitudinális vizsgálata* 53 (Longitudinal Research of Adolescent Mothers' Social and Demographic Features), 216.

Emil Valkovics

Cause of Death as a Factor in Creating Differences in Mortality Levels

The mortality level of the Hungarian male population has been declining since the mid 1960s. In the case of the Hungarian female population, we are witnessing a stagnation, accompained by very small declines and rises of the mortality level. The mortality level of Hungary's overall population, similar to the mortality levels of the populations of other East European countries, that were socialist in the past, is rising, i.e. deteriorating. The mortality level observed in these countries is, as is well known, much higher than the observed mortality levels in other European countries and in many non European countries. The analysis of differences between mortality levels is therefore a task of crucial importance for Hungarian demographers. We intend in this modest contribution to answer a question of a purely methodological nature: How can we measure correctly the differences in mortality levels and how can we explain the role of different causes of death in creating differences in mortality levels?

Demographers consider the best measure of the mortality level to be the life expectancy at birth, and the differences in mortality levels may be explained by analyzing the differences in life expectancies at birth. If the mortality level declines, the life expectancy at birth rises, and inversely: if the mortality level rises, the life expectancy at birth declines.

Supposing the knowledge and understanding of the life table functions and the relationships between them, we clarify more in detail only the meaning of the $_nL_x$ function of the life table because it may be considered as an immediate determinant of the life expectancy at birth ($e^0{}_0$). If the radix, i.e. the number of liveborn in the life table equals the unity ($l_0 = 1$), the life expectancy at birth may be considered as the sum of the values in the $_nL_x$ column of the life table, i.e.:

$$e^0_0 = \sum_{x=0}^{w} {}_nL_x = T_0,$$

where $_nL_x$ may be considered:

1. as the number of person-years to be lived by the newborn (l_0) between the exact ages x and $x+n$, T_0 being the total number of these years, i.e. the total after lifetime at the exact age of 0 years;

2. as the number of person-years lived by the deceased of the life table ($_nd_x$), the total number of these years equals also the life expectancy at birth;

3. as the number of person-years, if considered separately for each age-interval, lived partly by the survivors (l_x) partly by the deceased between the exact ages x and $x+n$:

$$_nL_x = nl_{x+n} + (\ _nL_x - nl_{x+n}),$$

where l_{x+n} denotes the number of survivors at the exact age $x+n$.

4. as the life table stationary population between the exact ages x and $x+n$, T_0 being in this case the total number of the life table stationary population exposed to the risk of dying during the period of time studied. The number of deceased ($_nd_x$) in the life table may be calculated by multiplying $_nL_x$ by $_nm_x$ (age specific death rate)

$$_nL_x\ _nm_x = \ _nd_x,$$

or

$$l_x\ _nq_x = \ _nd_x,$$

where l_x is the number of survivors at the exact age x and $_nq_x$ is the probability of dying between the exact ages x and $x+n$.

These meanings of the $_nL_x$ function are true in the case when we work with complete life tables, containing single year age intervals ($n=1$) and in the cases when we work with abridged life tables containing larger age groups ($n > 1$).

The first attempts to explain the rise or decline of the life expectancy at birth were based on the first meaning of the $_nL_x$ function. The authors who elaborated these explanations all underlined that the changes of the values of the $_nL_x$ function are originated in changes of the age-specific mortality rates ($_nm_x$), but only one of them, professor *John H. Pollard* began this explanation by showing the changing values of the $_nm_x$ function and its consequences (1982). The other succesful authors: professor *Eugeny M. Andreev* (1982) and professor *Roland Pressat* (1985, 1995) underlined also correctly the origin of the changes, but the changing values of the $_nm_x$ function (its total value and the values distributed by causes of death) entered in the field of calculations only in their last stage. Professor *Eduardo E. Arriaga* made also an interesting attempt to explain the change in life expectancies (1984).

Table 1 shows the number of surviving males ($l^{(M)}{}_x$) and females ($l^{(F)}{}_x$), the life expectancies of males ($e^{0(M)}{}_x$) and females ($e^{0(F)}{}_x$); the total after lifetimes of males ($T^{(M)}{}_x$) and of females ($T^{(F)}{}_x$), the number of person-years of males ($_nL^{(M)}{}_x$) and females ($_nL^{(F)}{}_x$) and the differences between the life expectancies at the exact age of 0, 20 and 60 years according to the abridged Hungarian life tables of Hungarian males and females for 1990. The relationships between these measures are well known:

$$T_x = l_x e_x^0 = \sum_{x=\omega}^{x} {}_nL_x$$

$$_nL_x = T_x - T_{x+n}$$

$$e_x^0 = T_x / l_x = \sum_{x=\omega}^{x} {}_nL_x / l_x$$

$$l_x = T_x / e_x^0 = \sum_{x=\omega}^{x} {}_nL_x / e_x^0$$

etc.

It is clear that the difference between the life expectancies at birth of Hungarian females and males $e^{0(F)}{}_0 - e^{0(M)}{}_0 = 73.71010 - 65.13000 = 8.58010$ years equals the sum of differences between the numbers of person-years lived by the females and males $0.08300 + 2.29291 + 6.20419 = 8.58010$ years as illustrated by the column (10) of the Table 1.

John H. Pollard, Eugeny M. Andreev and *Roland Pressat* calculate the same sum of differences, but not by creating and summarizing the differences between the values of $_nL^{(F)}{}_x$ and $_nL^{(M)}{}_x$ columns as shown in the column (10) of the Table 1. They distribute the differences by the age-groups of their origin. They utilize different formulae, but obtain entirely or approximately the same results (see the column (2) and the column (5) of the Table 2).

John H. Pollard evaluates the differences between life expectancies at birth and their distribution by causes of death by calculating

$$e_0^{0(F)} - e_0^{0(M)} = \sum_i ({}_1m_{i,0}^{(M)} - {}_1m_{i,0}^{(F)}) \, w_{0.5}$$

$$+ 4\sum_i ({}_4m_{i,1}^{(M)} - {}_4m_{i,1}^{(F)}) \, w_3$$

$$+ 5\sum_i ({}_5m_{i,5}^{(M)} - {}_5m_{i,5}^{(F)}) \, w_{7.5}$$

$$+ 5\sum_i ({}_5m_{i,10}^{(M)} - {}_5m_{i,10}^{(F)}) \, w_{12.5} + \ldots,$$

where

$$w_x = \frac{1}{2}(l_x^{(F)} e_x^{0(M)} + l_x^{(M)} e_x^{0(F)}), \text{ if } l_0 = 1.$$

Finally

$$e_0^{0(F)} - e_0^{0(M)} = \sum_x \sum_i (Q_{i,x}^{(M)} - Q_{i,x}^{(F)}) \, w_x,$$

where

$$Q_{i,x} = -\ln\left(\frac{l_{i,x+n}}{l_{i,x}}\right) = -\ln {}_nP_{i,x} = {}_nm_{i,x}$$

and $_nm_{i,x}$ denotes the age-specific death rates due to cause of death i, the general age-specific death rates being equal to the sums of age-specific death rates by causes of death.

In our modest contribution we use the subscript i for denoting the mortality due to circulatory diseases (including the cerebro-vascular diseases) and $\sim i$ for denoting the mortality due to all othes causes of death.

The abridged life table we use contains only three age groups, we cannot illustrate therefore the use of the original method of *John H. Pollard*, we note only that the results we obtain by using it are identical or almost identical with the results we obtain when using the method of *Eugeny M. Andreev*.

Eugeny M. Andreev derives and utilizes the formula

$$l_x^{(F)} [e_x^{0(F)} - e_x^{0(M)}] - l_{x+n}^{(F)} [e_{x+n}^{(F)} - e_{x+n}^{(M)}].$$

We obtain by using this formula e.g. for the age group of aged 20–59 years **0.98067** (**55.09561** − **46.78116**) − **0.85246** (**19.02015** − **14.71781**) = **4.48616** years.

Roland Pressat derives and utilizes the formula

$$\left[\frac{l_{x+n}^{(M)} + l_{x+n}^{(F)}}{2}(e_{x+n}^{0(M)} - e_{x+n}^{0(F)})\right] - \left[\frac{l_x^{(M)} + l_x^{(F)}}{2}(e_x^{0(M)} - e_x^{0(F)})\right].$$

We obtain by using his formula for the same age group

$$\left[\frac{0.68011 + 0.85246}{2} (14.71781 - 19.02015) \right] -$$

$$- \left[\frac{0.97333 + 0.98067}{2} (46.78116 - 55.09561) \right] = 4.82650.$$

The gains calculated by using these formulae are distributed by causes of death studied in all the causes by using the formulae

$$\frac{{}_n m^{(M)}_{i,x} - {}_n m^{(F)}_{i,x}}{{}_n m^{(M)}_x - {}_n m^{(F)}_x} \quad \text{and} \quad \frac{{}_n m^{(M)}_{\sim i,x} - {}_n m^{(F)}_{\sim i,x}}{{}_n m^{(M)}_x - {}_n m^{(F)}_x},$$

where ${}_n m^{(M)}_x$ and ${}_n m^{(F)}_x$ are the age-speci~fic death rates for males and females, ${}_n m^{(M)}_{i,x}$ and ${}_n m^{(F)}_{i,x}$ denote the age-specific death rates du to circulatory diseases according to proportions observed in Hungary in 1990 and ${}_n m^{(M)}_{\sim i,x}$ and ${}_n m^{(F)}_{\sim i,x}$ are the age-specific death rates due to all other causes of death (see the details in Table 4.1 and Table 4.2). Table 4.3 contains the corresponding probabilities of dying (${}_n q_x$, ${}_n q_{i,x}$ and ${}_n q_{\sim i,x}$).

It is clear on the basis of the data in the last line of Table 2, that from the difference between the life expectancies at birth of Hungarian males and Hungarian females (i.e. form 8.58010 years) only 2.41855 years are due to circulatory diseases and 6.16155 years to all other causes of death, if we use the method of *John H. Pollard* or *Eugeny M. Andreev* and 2.45365 years are due to circulatory diseases and 6.12645 to all the other causes of death, if we use the method of *Roland Pressat*. These results are somewhat different, but the differences between them are trifling.

Eugeny M. Andreev underlines that his method is based on the methodological results of *Korchak-Chepurkovskiy* published in 1968. This method is presented in Table 3.1 and Table 3.2. Table 3.1 shows how to calculate the direct and indirect contribution of mortality change in different age groups to the number of person-years lived in the same age groups and in other age groups. Column (1) of Table 3.1 enumerates the age groups, column (2) shows the ratios between the values of the survivorship funtions of Hungarian females and males obtained by division of the data of column (3) by the data of column (2) of Table 1. Column (3) of Table 3.1 repeats the values of the $_nL^{(M)}{}_x$ function of the life table for males (column (8) of Table 1).

In the first line of column (4) we introduce the value of the first line of the $_nL^{(F)}{}_x$ function of the life table for females (column (9) of Table 1). The other values in this column are obtained by multiplication of the data (in corresponding age groups of column (3) by 0.98067/0.97333 = = 1.99754.

Columm (5) denoted by Δ_1 contains the differences between the values in column (4) and column (3). The sum of these differences, i.e. 0.42632 is naturally equal to the difference between the sums of data in column (4) and column (3). The gain in the youngest age group equals simply the difference between the corresponding values of $_nL^{(F)}{}_x$ function and $_nL^{(M)}{}_x$ function (19.67949 − 19.59649 = 0.08300, as shown also in Table 1), i.e. in the case of the youngest age group we have only a direct contribution of mortality differences to the differences between the numbers of person-years (the multiplication of this contribution by $l^{(F)}{}_0/l^{(M)}{}_0 = 1/1 = 1$ does not influence the amount of this contribution).

The first two lines of column (6) of Table 3.1 contain the first two values taken from the $_nL^{(F)}{}_x$ function of the life table for females (column (9) of Table 1). The following value is obtained by multiplication of the corresponding figure in column (3) by 0.85246/0.68011 = 1.25341.

Column (7) denoted by the Δ_2 contains the differences between the values in column (6) and column (4). The sum of these differences, i.e. 4.48616 is naturally equal to the difference between the sums of data in column (6) and column (4).

Column (8) of Table 3.1 contains simply the direct contribution of mortality differences to the differences in numbers of person-years in the oldest age group.

Table 3.2 shows, the direct and indirect contribution of morality differences to the differences in numbers of person-years by age groups. Their sum is shown in column (4) of this table and is the same we obtained by using the method of *Eugeny M. Andreev.*

The direct contributions shown in the column (2) of Table 3.2 are not comparable because we work with different age intervals and these contributions are more important, if the age interval is longer and inversely.

Let us add to this objection that the age intervals we use influence the gains in numbers of person-years by causes of death studied too. If we use for the same comparison the abridged life tables containing the usual five year age intervals (0, 1–4, 5–9, 10–14, ..., 85+ years), the gains due to circulatory diseases will amount to 3.07630 years and the gains due to all other causes of death to 5.50380 years of we use the method of *Pollard* and *Andreev* (instead of 2,41855 years and 6,15155 years presented in the last line of Table 2). The age intervals influence naturally the total amount of direct and indirect contributions too. The direct contributions will amount only to 1,48367 years and the indirect contributions to 7,09643 years when using the usual five year age groups instead of results presented in Table 3.2. This is naturally a disadvantage of the methods in question.

We note that *Korchak-Chepurkovskiy* did not distribute the (direct and indirect) gains by causes of death. *Eugeny M. Andreev* did it, as it was already shown.

Table 5.1 and 5.2 show the results of an other type of calculations based on the data of Table 4.1 and Table 4.2. The survivors, the numbers of person-years, the total after lifetime and the life expectancies at births distributed by causes of death studied are presented in them.

Table 6 shows the structure of survivors at age x, i.e. the structure of deceased above the age x by causes of death studied.

Table 7 presents the life expectancies at age x as weighted arithmetic means and the differences between them due to causes of death studied. It is clear that the life expectancy at birth is a weighted arithmetic mean of life expectancies at birth of victims of causes of death studied.

The differences between the life expectancies at birth are equal to 8.58010 years as before, from which 12.710908 years are due to circulatory diseases and −4.13088 to all other causes of death. These results are very different from the results we obtained by using the method of *John H. Pollard* and *Eugeny M. Andreev* (2.41855 years and 6.16155 years) and from the results we obtained by using the method of *Roland Pressat* (2.45365 years and 6.12645 years). In principle it is possible to give only one correct answer to the question concerning the role of different causes of death in creating differences in life expectancies at birth. Which is from the answers we obtained correct and which is not correct? Are they really answers to the same question?

Eugeny M. Andreev, when introducing his method, decomposes the differences between total after lifetimes and the approach of *Roland Pressat* is also similar. The total after lifetimes are products of two other life table functions, the decomposition of their differences is relatively easy. *Andreev* shows that

$$T_x^{(F)} - T_x^{(M)} = l_x^{(F)} e_x^{0(F)} - l_x^{(M)} e_x^{0(M)} =$$

$$= (l_x^{(F)} - l_x^{(M)}) e_x^{0(M)} + l_x^{(M)} (e_x^{0(F)} - e_x^{0(M)}) +$$

$$+ (l_x^{(F)} - l_x^{(M)}) (e_x^{0(F)} - e_x^{0(M)}) =$$

$$= (l_x^{(F)} - l_x^{(M)}) e_x^{0(M)} + l_x^{(F)} (e_x^{0(F)} - e_x^{0(M)})$$

and he is undoubtedly right. He utilizes the second term of the formula containing only two terms: the

$$l_x^{(F)} (e_x^{0(F)} - e_x^{0(M)})$$

for calculating the amount of gains (the direct *and* indirect contributions of changing mortality levels in different age groups to the number of person-years) and their distribution according to the age groups of their origin.

Is it really reasonable to begin with the decomposition of differences between total after lifetimes?

Several life table functions may be considered also as products of two other life table functions. This fact enables us to use the well known and relatively simple mathematical formulae we use when decomposing the differences between two products as *Andreev* did it.

$$ab - cd = (a - c)d + (b - d)c + (a - c)(b - d) = (a - c)d + (b - d)a$$

or

$$ab - cd = (b - d)c + (a - c)d + (a - c)(b - d) = (a - c)b + (b - d)c.$$

It is clear that

$$(a - c)d + (b - d)a = (a - c)b + (b - d)c$$

on the basis of the evidence:

$$(a - c)(b - d) = (a - c)(b - d).$$

It is clear also that the product *ab* may be obtained among others if we add to the product *cd* the difference between the products *ab* and *cd*.

The differences consisting from two terms only $[(a - c)d + (b - d)a]$ and $[(a - c)b + (b - d)c]$ are elements of set of formulae we utilize when realizing a double standardization (*Kitagawa* 1955, 1964).

We decompose first the differences between the life table death functions: $_nd^{(F)}_x$ and $_nd^{(M)}_x$ in our case. The changing mortality level, the changing set of values of the $_nm_x$, $_nq_x$ or $_np_x$ functions leads always to the change of all other life table funcitons. In the case of $_nd_x$ function the change of the age-specific mortality rates leads to the change of the age distribution of deceased, the sum of deceased will remain always the same, it will remain always equal to the number of the new born in the life table.

$$(\sum_{\omega}^{0} {}_nd_x = l_0)$$

The life expectancy at birth will remain always the mean age of deceased in the life table and this mean age will remain always the mean age of the mean ages at death of victims of causes of death studied.

The decomposition of differences between $_nd^{(F)}_x$ and $_xd^{(M)}_x$ is shown in Table 8. The differences between $_nd^{(F)}_x$ and $_xd^{(M)}_x$ may be naturally decomposed not only as it is presented in two parts of Table 8. They may be decomposed also as it was already mentioned as follows:

$$_nd^{(F)}_x - {}_nd^{(M)}_x = {}_nL^{(F)}_x {}_nm^{(F)}_x - {}_nL^{(M)}_x {}_nm^{(M)}_x =$$

$$= l^{(F)}_x {}_nq^{(F)}_x - l^{(M)}_x {}_nq^{(M)}_x =$$

$$= {}_nL^{(M)}_x ({}_nm^{(F)}_x - {}_nm^{(M)}_x) + ({}_nL^{(F)}_x - {}_nL^{(M)}_x) {}_nm^{(M)}_x +$$

$$+ ({}_nL^{(F)}_x - {}_nL^{(M)}_x)({}_nm^{(F)}_x - {}_nm^{(M)}_x) =$$

$$= {}_nL^{(M)}_x ({}_nm^{(F)}_x - {}_nm^{(M)}_x) + ({}_nL^{(F)}_x - {}_nL^{(M)}_x) {}_nm^{(F)}_x =$$

$$= l^{(M)}_x ({}_nq^{(F)}_x - {}_nq^{(M)}_x) + (l^{(F)}_x - l^{(M)}_x) {}_nq^{(M)}_x +$$

$$+ (l^{(F)}_x - l^{(M)}_x)({}_nq^{(F)}_x - {}_nq^{(M)}_x) =$$

$$= l^{(M)}_x ({}_nq^{(F)}_x - {}_nq^{(M)}_x) + (l^{(F)}_x - l^{(M)}_x) {}_nq^{(F)}_x.$$

It is clear that

$$_nd_x^{(F)} - {_nd_x^{(M)}} =$$
$$= ({_nL_x^{(F)}} - {_nL_x^{(M)}}) {_nm_x^{(M)}} + {_nL_x^{(F)}} ({_nm_x^{(F)}} - {_nm_x^{(M)}}) =$$
$$= ({_nL_x^{(F)}} - {_nL_x^{(M)}}) {_nm_x^{(F)}} + {_nL_x^{(M)}} ({_nm_x^{(F)}} + {_nm_x^{(M)}}) =$$
$$= (l_x^{(F)} - l_x^{(M)}) {_nq_x^{(M)}} + l_x^{(F)} ({_nq_x^{(F)}} - {_nq_x^{(M)}}) =$$
$$= (l_x^{(F)} - l_x^{(M)}) {_nq_x^{(F)}} + l_x^{(M)} ({_nq_x^{(F)}} - {_nq_x^{(M)}})$$

and

$$({_nL_x^{(F)}} - {_nL_x^{(M)}}) {_nm_x^{(F)}} - ({_nL_x^{(F)}} - {_nL_x^{(M)}}) {_nm_x^{(M)}} =$$
$$= {_nL_x^{(F)}} ({_nm_x^{(F)}} - {_nm_x^{(M)}}) - {_nL_x^{(M)}} ({_nm_x^{(F)}} - {_nm_x^{(M)}})$$

because

$$({_nL_x^{(F)}} - {_nL_x^{(M)}})({_nm_x^{(F)}} - {_nm_x^{(M)}}) =$$
$$= ({_nL_x^{(F)}} - {_nL_x^{(M)}})({_nm_x^{(F)}} - {_nm_x^{(M)}})$$

and

$$(l_x^{(F)} - l_x^{(M)}) {_nq_x^{(F)}} - (l_x^{(F)} - l_x^{(M)}) {_nq_x^{(M)}} =$$
$$= l_x^{(F)} ({_nq_x^{(F)}} - {_nq_x^{(M)}}) - l_x^{(M)} ({_nq_x^{(F)}} - {_nq_x^{(M)}})$$

because

$$(l_x^{(F)} - l_x^{(M)})({_nq_x^{(F)}} - {_nq_x^{(M)}}) = (l_x^{(F)} - l_x^{(M)})({_nq_x^{(F)}} - {_nq_x^{(M)}}).$$

Table 9 presents the decomposition between $l^{(F)}_x$ and $l^{(M)}_x$. The differences between $l^{(F)}_x$ and $l^{(M)}_x$ may be also decomposed not only as it is presented in two parts of Table 9. They may be decomposed in the following way too

$$l^{(F)}_x - l^{(M)}_x = \sum_{\omega}^{x} {}_nL^{(F)}_x \, {}_nm^{(F)}_x - \sum_{\omega}^{x} {}_nL^{(M)}_x \, {}_nm^{(M)}_x =$$

$$= \sum_{\omega}^{x} l^{(F)}_x \, {}_nq^{(F)}_x - \sum_{\omega}^{x} l^{(M)}_x \, {}_nq^{(M)}_x =$$

$$= \sum_{\omega}^{x} {}_nL^{(M)}_x ({}_nm^{(F)}_x - {}_nm^{(M)}_x) +$$

$$+ \sum_{\omega}^{x} ({}_nL^{(F)}_x - {}_nL^{(M)}_x) \, {}_nm^{(M)}_x +$$

$$+ \sum_{\omega}^{x} ({}_nL^{(F)}_x - {}_nL^{(M)}_x)({}_nm^{(F)}_x - {}_nm^{(M)}_x) =$$

$$= \sum_{\omega}^{x} {}_nL^{(M)}_x ({}_nm^{(F)}_x - {}_nm^{(M)}_x) +$$

$$+ \sum_{\omega}^{x} ({}_nL^{(F)}_x - {}_nL^{(M)}_x) \, {}_nm^{(F)}_x =$$

$$= \sum_{\omega}^{x} l^{(M)}_x ({}_nq^{(F)}_x - {}_nq^{(M)}_x) + \sum_{\omega}^{x} (l^{(F)}_x - l^{(M)}_x) \, {}_nq^{(M)}_x +$$

$$+ \sum_{\omega}^{x} (l^{(F)}_x - l^{(M)}_x)({}_nq^{(F)}_x - {}_nq^{(M)}_x) =$$

$$= \sum_{\omega}^{x} l^{(M)}_x ({}_nq^{(F)}_x - {}_nq^{(M)}_x) + \sum_{\omega}^{x} (l^{(F)}_x - l^{(M)}_x) \, {}_nq^{(F)}_x .$$

It is clear that

$$l_x^{(F)} - l_x^{(M)} =$$

$$= \sum_{\omega}^{x} ({}_nL_x^{(F)} - {}_nL_x^{(M)}) \, {}_nm_x^{(M)} +$$

$$+ \sum_{\omega}^{x} {}_nL_x^{(F)} ({}_nm_x^{(F)} - {}_nm_x^{(M)}) =$$

$$= \sum_{\omega}^{x} ({}_nL_x^{(F)} - {}_nL_x^{(M)}) \, {}_nm_x^{(F)} +$$

$$+ \sum_{\omega}^{x} {}_nL_x^{(M)} ({}_nm_x^{(F)} + {}_nm_x^{(M)}) =$$

$$= \sum_{\omega}^{x} (l_x^{(F)} - l_x^{(M)}) \, {}_nq_x^{(M)} + \sum_{\omega}^{x} l_x^{(F)} ({}_nq_x^{(F)} - {}_nq_x^{(M)}) =$$

$$= \sum_{\omega}^{x} (l_x^{(F)} - l_x^{(M)}) \, {}_nq_x^{(F)} + \sum_{\omega}^{x} l_x^{(M)} \sum_{\omega}^{x} ({}_nq_x^{(F)} - {}_nq_x^{(M)})$$

and

$$\sum_{\omega}^{x} ({}_nL_x^{(F)} - {}_nL_x^{(M)}) \, {}_nm_x^{(F)} - \sum_{\omega}^{x} ({}_nL_x^{(F)} - {}_nL_x^{(M)}) \, {}_nm_x^{(M)} =$$

$$= \sum_{\omega}^{x} {}_nL_x^{(F)} ({}_nm_x^{(F)} - {}_nm_x^{(M)}) = \sum_{\omega}^{x} {}_nL_x^{(M)} ({}_nm_x^{(F)} - {}_nm_x^{(M)})$$

because

$$\sum_{\omega}^{x}(\ _nL_x^{(F)}-\ _nL_x^{(M)})(\ _nm_x^{(F)}-\ _nm_x^{(M)})=$$

$$=\sum_{\omega}^{x}(\ _nL_x^{(F)}-\ _nL_x^{(M)})(\ _nm_x^{(F)}-\ _nm_x^{(M)})$$

and

$$=\sum_{\omega}^{x}(l_x^{(F)}-l_x^{(M)})\ _nq_x^{(F)}-\sum_{\omega}^{x}(l_x^{(F)}-l_x^{(M)})\ _nq_x^{(F)}-$$

$$-\sum_{\omega}^{x}(l_x^{(F)}-l_x^{(M)})\ _nq_x^{(M)}=$$

$$=\sum_{\omega}^{x}l_x^{(F)}(\ _nq_x^{(F)}-\ _nq_x^{(M)})-\sum_{\omega}^{x}l_x^{(M)}(\ _nq_x^{(F)}-\ _nq_x^{(M)})$$

because

$$\sum_{\omega}^{x}(l_x^{(F)}-l_x^{(M)})(\ _nq_x^{(F)}-\ _nq_x^{(M)})=$$

$$=\sum_{\omega}^{x}(l_x^{(F)}-l_x^{(M)})(\ _nq_x^{(F)}-\ _nq_x^{(M)}).$$

Eugeny M. Andreev starts from the first of the mathematical evidence we use when decomposing the difference between two products

$$T_x^{(F)} - T_x^{(M)} = l_x^{(F)} e_x^{0(F)} - l_x^{(M)} e_x^{0(M)} =$$
$$= (l_x^{(F)} - l_x^{(M)}) e_x^{0(M)} + l_x^{(M)} (e_x^{0(F)} - e_x^{0(M)}) +$$
$$+ (l_x^{(F)} - l_x^{(M)})(e_x^{0(F)} - e_x^{0(M)}) =$$
$$= (l_x^{(F)} - l_x^{(M)}) e_x^{0(M)} + l_x^{(F)} (e_x^{0(F)} - e_x^{0(M)})$$

which is undoubtedly correct. The difference decomposed only in two terms is known, as it was already mentioned, as one of the set of formulae of double standardization. Nevertheless the above difference may be decomposed also as follows

$$T_x^{(F)} - T_x^{(M)} = l_x^{(F)} e_x^{0(F)} - l_x^{(M)} e_x^{0(M)} =$$
$$= l_x^{(M)} (e_x^{0(F)} - e_x^{0(M)}) + (l_x^{(F)} - l_x^{(M)}) e_x^{0(M)} +$$
$$+ (l_x^{(F)} - l_x^{(M)})(e_x^{0(F)} - e_x^{0(M)}) =$$
$$= (l_x^{(F)} - l_x^{(M)}) e_x^{0(F)} + l_x^{(M)} (e_x^{0(F)} - e_x^{0(M)})$$

which belongs naturally also to the sets of formulae of double standardization.

The values of

$$l_x^{(M)} (e_x^{0(F)} - e_x^{0(M)})$$

are given in the column (4) of Table 10. If we add to them the values of

$$(l_x^{(F)} - l_x^{(M)}) e_x^{0(F)}$$

we obtain also

$$T_x^{(F)} - T_x^{(M)}.$$

Age (years) x	$l_x^{(M)}(e_x^{0(F)}-e_x^{0(M)})$	$(l_x^{(F)}-l_x^{(M)})e_x^{0(F)}$	$T_x^{(F)}-T_x^{(M)}$
(1)	(2)	(3)	(4)=(2)+(3)
0	8.58010	0.00000	8.58010
20	8.09270	0.40440	8.49710
60	2.92606	3.27812	6.20419

For the differences of $_nL_x$ function, shown in Table 11, we obtain

Age groups (years) x, x+n	$l_x^{(M)}(e_x^{0(F)}-e_x^{(M)})-$ $-l_{x+n}^{(M)}(e_{x+n}^{0(F)}-e_{x+n}^{(M)})$	$(l_x^{(F)}-l_x^{(M)})e_x^{0(F)}-$ $-(l_{x+n}^{(F)}-l_{x+n}^{(M)})e_{x+n}^{0(F)}$	$L_x^{(F)}-L_x^{(M)}$
(1)	(2)	(3)	(4)=(2)+(3)
0–19	0.48740	-0.40440	0.08300
20–59	5.16664	-2.87372	2.29292
60–	2.92600	3.27812	6.20418
Σ	8.58010	0.00000	8.58010

$$[(l_x^{(F)}-l_x^{M})e_x^{0(M)}-(l_{x+n}^{(F)}-l_{x+n}^{(M)})e_{x+n}^{0(M)}] \neq$$
$$\neq [(l_x^{(F)}-l_x^{(M)})(e_x^{0(F)}-e_x^{0(M)})-$$
$$-(l_{x+n}^{(F)}-l_{x+n}^{(M)})(e_{x+n}^{0(F)}-e_{x+n}^{0(M)})]$$

and these terms are not equal to

$$[(l_x^{(F)}-l_x^{M})e_x^{0(F)}-(l_{x+n}^{(F)}-l_{x+n}^{(M)})e_{x+n}^{0(F)}],$$

but their sums are equal to 0 in all the cases.

Table 9 showing the decomposition of differences between the survivorship functions may be used for producing more details for the

$$(l_x^{(F)} - l_x^{(M)}) e_x^{0(F)}$$

function.

It is clear that

$$T_x^{(F)} - T_x^{(M)} = (l_x^{(F)} - l_x^{(M)}) e_x^{(M)} + (e_x^{0(F)} - e_x^{0(M)}) l_x^{(F)} =$$
$$= (l_x^{(F)} - l_x^{(M)}) e_x^{0(F)} + (e_x^{0(F)} - e_x^{0(M)}) l_x^{(M)}$$

and

$$(l_x^{(F)} - l_x^{(M)}) e_x^{(F)} - (l_x^{(F)} - l_x^{(M)}) e_x^{0(M)} =$$
$$= (e_x^{0(F)} - e_x^{0(M)}) l_x^{(F)} - (e_x^{0(F)} - e_x^{0(M)}) l_x^{(M)}$$

because

$$(l_x^{(F)} - l_x^{(M)})(e_x^{0(F)} - e_x^{0(M)}) = (l_x^{(F)} - l_x^{(M)})(e_x^{0(F)} - e_x^{0(M)}).$$

It is clear in any case that it is not sufficient to utilize only one component, if we want to reproduce the differences between the values of the life table functions we compare. It is necessary to take into account the other components creating the differences too for obtaining such a distribution of differences which we obtain by calculating simply the differences between the values of corresponding functions. If we distribute the differences by using the methods of enumerated authors we will not be able to produce correctly for example the values of the $_nd^{(F)}_x$ function by multiplying the values of $_nL^{(F)}_x$ by $_nm^{(F)}_x$, or the values of $l^{(F)}_x$ by $_nq^{(F)}_x$. Not only the distribution of deceased will be quite unrealistic, but the total number of deceased in the life table will also deviate form the radix of the table (l_0), the equality

$$\sum_{\omega}^{0} {}_nd_x = l_0$$

will not be respected at all and the values of the other life table functions calculated from the values of ${}_nd^{(F)}_x$ function will be quite absurd too. The number of years lived by the deceased of the life table will not be equal to the sum of the values of the ${}_nL_x$ function. It will be impossible to demonstrate that the life expectancy at birth equals the mean age of all the deceased of the life table and this mean age is the weighted mean of mean ages of victims of different causes of death etc.

Does that mean that the presented methods are not useful at all and we must refuse them entirely? The answer for this question is rather complicated. Everybody who is alive and is exposed actually to risk of dying survived due to the mortality conditions in the past, at his (or her) youngest ages which made possible for him (or her) to survive. If we consider the correctly calculated ${}_nL_x$ function of the life table as the number of persons exposed to risk of dying between the exact ages x and $x+n$ and calculate the number of deceased by using the formulae ${}_nd_x = {}_nL_x \, {}_nm_x$ or ${}_nd_x = l_x \, {}_nq_x$, we obtain such a final sum of deceased which is equal to the number of newborn in the life table:

$$\sum_{x=\omega}^{0} {}_nd_x = l_0$$

Table 3.2 presented the direct and indirect contribution of mortality differences to the differences in numbers of person-years calculated by using the method of *Korchak-Chepurkovskiy* and *Eugeny M. Andreev*. The column (4) summarizes the contributions and distributes them according to their origin. If we change the data in column (3), showing the indirect contributions by origin and put this gains in numbers of person-years in those age groups in which they may be interpreted as persons exposed to the risk of dying, we obtain such a table, in which the sums of contributions will equal simply the differences between the values of ${}_nL^{(F)}_x$ column and ${}_nL^{(M)}_x$ column. These differences plus the values of the ${}_nL^{(M)}_x$ column equal the values of the ${}_nL^{(F)}_x$ column, which multiplied by the values of the ${}_nm^{(F)}_x$ column gives us the values of the ${}_nd^{(F)}_x$ column and of other measures of the life table for females.

On the basis of Table 3.3 both approaches are respected and the role of both sets of causes of death may be demonstrated.

We note nevertheless that a life table for females may be not considered as a transformed version of life table for males and inversely. It is somewhat more natural to compare two life tables by using the presented methods for the same sex but different years of lareger periods of time. We accept the presented procedures mainly on purely technical basis.

If we compare two life tables for the same sex, it is not easy at all to understand how the changing mortality at younger ages influenced the number of persons exposed to risk of dying at older ages if the difference in time between the life tables is short for that. Everybody needs e.g. 70 years for surviving from the age of 0 years till the age of 70 years and the difference in time between the life table e.g. for 1960 and 1990 is only 30 years.

Let us return to the method of decomposition of differences between the life expectancies at the age x elaborated and used in the *Demographic Research Institute of the HCSO*, presented already briefly in Table 5.1, Table 5.2, Table 6 and Table 7. Let us repeat this example by working now with nine groups of causes of death and a residual group for all other causes of death. The diseases of the circulatory system and the cerebrovascular diseases are separated too in the Table 12 showing the decomposition of differences between the life expectancies at birth of Hungarian females and Hungarian males.

Table 12 presents the contribution of mortality due to causes of death studied to the differences between the life expectancies at birth. The general life expectancy at birth ($e^0{}_0$) is considered as a weighted arithmetic mean of life expectancies at birth of victims of different causes of death, with the other words: the mean age at death of all the deceased in the life table is a weighted arithmetic mean of mean ages at death of victims of different causes of death and shows the overall contribution of mortality due to groups of causes of death studied to the differences in question.

The method we use actually for decomposing the differences between the life expectancies may be used naturally 'mutatis mutandis' when decomposing the differences between the life expectancies of other populations, too.

References

Andreev, Eugeny M.: Method komponent v analize prichin smerti. (Component method applied to life expectancy analysis.) *Vestnik Statistiki (Herald of Statistics)* 1982, no 9: 42–47. (In Russian)

Arriaga, Eduardo, E.: Measuring and explaining the change in life expectancies. *Demography*, Volume 21, Number 1, 1984, pp. 83–96.

Dr. Barsy, Gy.: Halandósági táblák. *In:* Szerk.: *dr. Szabady Egon*: Bevezetés a demográfiába. *Közgazdasági és Jogi Könyvkiadó*, Budapest, 1964. pp. 371–392. 610 (Mortality Tables, in Introduction to Demography, edited by Szabady, Egon)

Benjamin, B.–Pollard, J.H.: Mortality and other Actuarial Statistics. *Heinemann*, 1980.

Benjamin, B.–Haycocks, H.W.: The Analysis of Mortality and other Actuarial Statistics. Cambridge, *Cambridge University Press*, p. 392

Bourgeous-Pichat, J.: Les limites de la démographie potentielle. *Revue de l'Institut International de Statistique*, 1951. no 1., pp. 13–27.

Brass, W.–Hill, K.: Estimating adult mortality from orphanhood. *Proceedings of the International Population Conference*, Volume 3, I.U.S.S.P., Liège, 1973, pp. 111–123.

Coale, A.J.–Demény, P.: Regional Model Life Tables and Stable Populations. *Princeton Universtity Press*, Princeton, N.Y., 1966, p. 872

Duchêne, J.: Ajustement des tables de mortalité du moment par les lois de Gompertz et de Makeham. Un essai de Comparaison. *Population et Famille*, 37, (1976–1), pp. 37–75.

Gärtner, K.: Sterblichkeit nach dem Familienstand. *Zeitschrift für Bevölkerungswissenschaft*, 16, 1990, 1, pp. 53–66.

Goldman, N.–Lord, G.: A new look at entropy and the life table. *Demography*, 23(2), 1986, pp. 275–282.

Henry, L.: Mesure indirecte de la mortalité des adultes. *Population*, 1960. no 3., pp. 457–466.

Henry, L.: Démographie. Analyse et modèles. *Librairie Larousse*, 1972, pp. 239–241. 341

Hersch, L.: De la démographie actuelle à la démographie potentielle. *Librairie de l'Université*, Genève, 1944.

Holzer, J.Z.: Demografia. *P.W.E.* Warszawa, 1970, p. 350

Höhn, Ch.–Pollard, J.H.: Persönliche Gewohnheiten und Verhaltensweisen und Steiblichkeits - unterschiede nach dem Familienstand in Bundesrepublic Deutschland. *Zeitschrift für Bevölkerungswissenschaft,* 18. Jahrgang Heft 4/1992, pp. 415–433.

Keyfitz, N.–Flieger, W.: World Population. An Analysis of Vital Data. *The Chicago University Press,* Chicago, 1968, pp. 9–10. 672

Keyfitz, N.: Introduction to the Mathematics of Population. Reading: *Addison-Wesley Publ. Co.*, 1968. p. 450

Keyfitz, N.–Golini, A.: Mortality Comparisons: The Male-Female Ratio. *Genus.* Vol. XXXI, No 1–4, 1975, pp. 1–34.

Kitagawa, E.M.: Components of a difference between two rates. *Journal of the American Statistical Association,* 1955.

Kitagawa, E.M.: Standardized Comparisons in Population Research. *Demography,* 1964, 1(1), pp. 196–315.

Korchak-Chepurkovskiy, Yuriy A.: Vliyanie smertnosti v raznyikh vozrastakh na uvelichenie sredney prodolzhitel' nosti zhizni. (The influence of mortality at different ages on average life expectancy.) In: Vosproizvodstvo naseleniya SSSR (Reproduction of the population USSR), Moscow, 1968 (In Russian)

Le Bras, H.: Lois de mortalité et âge limite. *Population,* 1976. no 3., pp. 655–692.

Ledermann, S.: Nouvelles tables-types de mortalité, *I.N.E.D., Travaux et Documents* Cahier No 53, Paris, 1969, p. 260 + 13 annexes.

Lyerly, S.B.: The contraharmonic mean. *American Statistician,* 28 (4). pp. 162–163.

Manton, K.G.–Stallard, E.: Recent trends in Mortality Analysis. *Orlando: Academic Press* 342

Mitra, S.: A short note on the Taeuber paradox. *Demography,* 15(4), 1978, pp. 621–624.

Page, H.J.–Wunsch, G.: Parental Survival Data: Some Results of the Application of Ledermann's Model Life Tables. *Population Studies,* 1976. p. no 1., 59–76.

Pallós, E.: Magyarország halandósági táblái 1900/01-től 1967/68-ig. A KSH Népességtudományi Kutató Intézet Közleményei 34. sz. Budapest, 1971, p. 220. Mortality tables of Hungary, 1900/1-1967/68. Publications of the Population Research Institute of the Central Statistical Office. No.34 Budapest, 1971, p. 220)

Pallós, E.: Magyarország népességének 1969–1970. évi halandósági táblája. p. 6 (Manuscript).

Pollard, J.H.: The expectation of life and its relationship to mortality. *Journal of the Institute of Actuaries*, 109(2), 1982, pp. 225–240.

Pollard, J.H.: On the decomposition of Changes in Expectation of Life and Differentials in Life Expectancy. *Demography*, Vol. 25, no. 2, May 1988, pp. 265–276.

Pressat, R.: L'analyse démographique. Méthodes, résultats, applications. Paris, *P.U.F.*, 1961, pp. 98–136. 402

Pressat, R.: Principes d'analyse. *Éditions de l'I.N.E.D.*, 1966, pp. 17–26. 153

Pressat, R.: L'analyse démographique. Méthodes, résultats, applications. Deuxième édition entièrement refondue. *P.U.F.* Paris, 1969, pp. 17–27. 321.

Pressat, R.: Contribution de écarts de mortailté par âge à la différence des vies moyennes. *Population*, 1985, n° 4–5, pp. 765–780.

Pressat, R.: Elements de démographie mathématique, *AIDELF*, Paris, 1995, pp. 22–26, 279.

Ryder, N.B.: Notes on Stationary Populations. *Population Index*, 1975. no 1. (Vol. 41, No. 1), pp. 3–28.

Tekse, K.: Bevezetés a stabil népesség elméletébe. *Statisztikai Kiadó Vállalat*, Budapest, 1975, p. 224 (Introduction to the Theory of, Stable Population, Statisztika Publishing House, Budapest, 1975, p. 224)

Theiss, E.: A népesedés mechanizmusának vizsgálata. *In*: Szerk.: dr. Szabady Egon: Bevezetés a demográfiába, *Közgazdasági és Jogi Könyvkiadó*, Budapest, 1964, pp. 480–499. 610 (Examination of the mechanism of population In: editor Szabady, Egon: Introduction to Demography, Közgazdasági and Jogi Publishing House, Budapest, 1964, pp. 480–499, 610)

Valkovics, E.: Gazdaságdemográfiai módszerek (Methods of Economic Demography). *Tankönyvkiadó*, Budapest, 1973, pp. 37–63. p. 482

Valkovics, E.: Konieczność rozbudowy systemu paratmetrów tablic trwania życia w zwiazku z potrzebami analiz spoleczo-ekonomicznych. *In*: Monografie i opracowania, n° 73/1, Warszawa, p. 56–72.

Valkovics, E.: A gazdaságdemográfiai elemzés elvei és módszerei. *Tankönyvkiadó.* Budapest, 1976, pp. 33–89. 378 (Principles and Methods of Econo-Demographic Analysis, Tankönyvkiadó Publishing House Budapest, 1976, pp. 33-89, 378)

Valkovics, E.: Utilisation des principes et des méthodes de l'analyse démographique dans l'analyse économique. *In*: L'analyse démographique et tes applications. *Éditions du CNRS*, Paris, 1977, pp. 375–391.

Valkovics, E.: L'évolution récente de la mortalité dans les pays de l'Est: Essai d'explication à partir de l'exemple hongrois. *Espace-Populations-Sociétés*, 1984, III, pp. 141–168.

Valkovics, E.: An attempt of decomposition of the differences between the life expectancies at age x (On the basis of abridged Italian life tables of 1972 and 1982). Working Paper 12/90, Dicembre 1990, *I.R.P.* Roma, p. 42

Valkovics, E.: An attempt of decomposition of the differences between the life expectancies at the age x on the basis of Belgian and Hungarian abridged life tables by causes of death. *C.B.G.S*, Werkdocument nr. 71, 1991. p. 47

Valkovics, E.: Development of mortality in Hungary. Past and recent trends. Budapest, p. 36 p. (Manscript)

Vallin, J.: La mortalité par génération en France, depuis 1899, *I.N.E.D. Travaux et Documents, Cahier No 63, P.U.F.*, Paris, 1973, p. 483

Vaupel, J.W.: How change in age-specific mortality affects life expectancy. *Population Studies*, 40(1), 1986, pp. 147–157.

Vielrose, E.: Zarys demografii potencjalnej. *P.W.N.*, Warszawa, 1958, p. 252

Wunsch, G.: Une méthode d'estimation de tables intercensitaires de mortalité régionales. *Population et Famille*, 1972, No 26–27, pp. 109–118.

Wunsch, G.: Le calcul des années vécues. Problèmes de cohérence dans l'établissement des tables de mortalité. *Population et Famille*, 50–51, 1980, pp. 107–117.

Wunsch, G.: The Life Table: an Overview. *Manuscript.* Presented at the EAPS Conference: Life Tables in Europe: Data, Methods, and Models. Louvain-la-Neuve, April 21–23, 1994, p. 18

Table 1

The survivors, life expectancies, total after lifetimes, number of person-years and differences between the life expectancies at the age x according to the abridged Hungarian life tables for males and females (1990)

Age (years) x	$l^{(M)}_x$	$l^{(F)}_x$	$e^{0(M)}_x$	$e^{0(F)}_x$	$T^{(M)}_x$	$T^{(F)}_x$	$_nL^{(M)}_x$	$_nL^{(F)}_x$	$_nL^{(F)}_x - _nL^{(M)}_x$	$e^{0(F)}_x - e^{0(M)}_x$
(1)	(2)	(3)	(4)	(5)	(6)	(7)	(8)	(9)	(10)	(11)
0	1.00000	1.00000	65.13000	73.71010	65.13000	73.71010	19.59645	19.67949	0.08300	8.58010
20	0.97333	0.98067	46.78116	55.09561	45.53351	54.03061	35.52378	37.81669	2.29291	8.31445
60	0.68011	0.85246	14.71781	19.02015	10.00973	16.21392	10.00973	16.21392	6.20419	4.30234
Σ	–	–	–	–	–	–	65.13000	73.71010	8.58010	–

Table 2
Gains in numbers of person-years distributed by the age-groups of their origin and by causes of death studied

Age groups (years) $x, x+n$	All gains	Gains due to		All gains	Gains due to	
		circulatory diseases	all other causes of death		circulatory diseases	all other causes of death
	calculated by using the method of Pollard and Andreev			calculated by using the method of Pressat		
(1)	(2)	(3)	(4)	(5)	(6)	(7)
0–19	0.42637	0.01122	0.41515	0.45688	0.01202	0.44486
20–59	4.48616	1.56923	2.91693	4.82650	1.68828	3.13822
60–	3.66757	0.83810	2.82947	3.29672	0.75335	2.54337
Σ	8.58010	2.41855	6.16155	8.58010	2.45365	6.12645

Table 3.1

Calculation of direct and indirect contributions of mortality differences to the differences in numbers of person-years (According to Korchak-Chepurkovskiy)

Age (years) x	$l^{(F)}_x / l^{(M)}_x$			$\Delta_1 = (4)-(3)$		$\Delta_2 = (6)-(4)$	Δ_3
(1)	(2)	(3)	(4)	(5)	(6)	(7)	(8)
0	1.00000	19.59649	19.67949	0.08300	19.67949	–	–
20	1.00754	35.52378	35.79163	0.26785	37.81669	2.02506	–
60	1.25341	10.00973	10.08520	0.07547	12.54630	2.46110	3.66762
Σ	–	65.13000	65.55632	0.42632*	70.04248	4.48616**	3.66762

* 0.42632 = 65.55632 − 65.13000
** 0.48616 = 70.04248 − 65.55632

Table 3.2
Direct and indirect contribution of mortality differences to the differences in numbers of person-years

Age groups (years) $x, x+n$	Direct contribution $_nL^{(F)}_x - _nL^{(M)}_x (l^{(F)}_x - l^{(M)}_x)$	Indirect contribution	Sums of contributions
(1)	(2)	(3)	(4)
0–19	0.08300	0.34337	0.42637
20–59	2.02506	2.46110	4.48616
60–	3.66762	–	3.66757
Σ	5.77568	2.80442	8.58010

Table 3.3
Direct and indirect contributions of mortality differences to the number of differences in number of person-years and other life tables measures

Age groups (years) $x, x+n$	Direct contribution $_nL_x^{(F)} - {_nL_x^{(M)}}(l_x^{(F)}/l_x^{(M)})$	Indirect contribution (rearranged)	Sums of contributions $_nL_x^{(F)} - {_nL_x^{(M)}}$	$_nL_x^{(M)}$	$_nm_x^{(F)}$	$_nd_x^{(F)}$	$l_x^{(F)}$	$T_x^{(F)}$	$e_x^{0(F)}$
(1)	(2)	(3)	(4)	(5)	(6)	(7)	(8)	(9)	(10)
0–19	0.08300	0.00000	0.08300	19.59645	0.00098	0.01933	1.00000	73.71010	73.71010
20–59	2.02506	0.26785	2.29291	35.52378	0.00339	0.12821	0.98067	54.03061	55.09561
60–	3.66762	2.53657	6.20419	10.00973	0.05258	0.85246	0.85246	16.21392	19.02015
Σ	5.77568	2.80442	8.58010	65.13000	–	1.00000	–	–	–

Table 4.1

Structure and number of deceased and of age-specific death rates by two groups of causes of death studied according to the abridged Hungarian life table for males (1990)

Age groups (years) $x, x+n$	Proportion of victims of			Number of victims of			Age-specific mortality rates		
	all causes of death	circulatory diseases	all other causes of death	all causes of death $_nd^{(M)}{}_x$	circulatory diseases $_nd^{(M)}{}_{i,x}$	all other causes of death $_nd^{(M)}{}_{\neg i,x}$	all causes of death $_nm^{(M)}{}_x$	circulatory diseases $_nm^{(M)}{}_{i,x}$	all other causes of death $_nm^{(M)}{}_{\neg i,x}$
(1)	(2)	(3)	(4)	(5)	(6)	(7)	(8)	(9)	(10)
0–19	1.00000	0.01912	0.98088	0.02667	0.00051	0.02616	0.00136	0.00003	0.00133
20–59	1.00000	0.32907	0.67093	0.29322	0.09649	0.19673	0.00825	0.00271	0.00554
60–	1.00000	0.55173	0.44827	0.68011	0.37524	0.30487	0.06794	0.03748	0.03046
Σ	–	–	–	1.00000	0.47224	0.52776	–	–	–

Table 4.2
Structure and number of deceased and of age-specific death rates by two groups of causes of death studied according to the abridged Hungarian life table for females (1990)

Age groups (years) $x, x+n$	Proportion of victims of			Number of victims of			Age-specific mortality rates		
	all causes of death	circulatory diseases	all other causes of death	all causes of death $_nd^{(F)}_x$	circulatory diseases $_nd^{(F)}_{i,x}$	all other causes of death $_nd^{(F)}_{\neg i,x}$	all causes of death $_nm^{(F)}_x$	circulatory diseases $_nm^{(F)}_{i,x}$	all other causes of death $_nm^{(F)}_{\neg i,x}$
(1)	(2)	(3)	(4)	(5)	(6)	(7)	(8)	(9)	(10)
0–19	1.00000	0.01811	0.98189	0.01933	0.00035	0.01898	0.00098	0.00002	0.00096
20–59	1.00000	0.29647	0.70353	0.12821	0.03801	0.09020	0.00339	0.00101	0.00238
60–	1.00000	0.64598	0.35402	0.85246	0.55067	0.30179	0.05258	0.03397	0.01861
Σ	–	–	–	1.00000	0.58903	0.41097	–	–	–

Table 4.3
The non independent probabilities of dying corresponding to Table 4.1 and 4.2

Age groups (years) $x, x+n$	$_nq_x^{(M)}$	$_nq_{i,x}^{(M)}$	$_nq_{\sim i,x}^{(M)}$	$_nq_x^{(F)}$	$_nq_{i,x}^{(F)}$	$_nq_{\sim i,x}^{(F)}$
(1)	(2)	(3)	(4)	(5)	(6)	(7)
0–19	0.02667	0.00051	0.02616	0.01933	0.00035	0.01898
20–59	0.30125	0.09913	0.20212	0.13074	0.03876	0.09198
60–	1.00000	0.55173	0.44827	1.00000	0.64598	0.35402

Table 5.1

The survivors, number of person-years, total after lifetime and life expectancies at the age x distributed by causes of death studied according to the abridged Hungarian life table by causes of death for males (1990)

Age (years) x	$l^{(M)}_{i,x}$	$l^{(M)}_{\neg i,x}$	$l^{(M)}_x$	$_nL^{(M)}_{i,x}$	$_nL^{(M)}_{\neg i,x}$	$_nL^{(M)}_x$	$T^{(M)}_{i,x}$	$T^{(M)}_{\neg i,x}$	$T^{(M)}_x$	$e^{0(M)}_{i,x}$	$e^{0(M)}_{\neg i,x}$	$e^{0(M)}_x$
(1)	(2)	(3)	(4)	(5)	(6)	(7)	(8)	(9)	(10)	(11)	(12)	(13)
0	0.47224	0.52776	1.00000	9.44040	10.15609	19.59649	33.37225	31.75775	65.13000	70.66799	60.17461	65.13000
20	0.47173	0.50160	0.97333	18.00224	17.52154	35.52378	23.93185	21.60166	45.53351	50.73209	43.06551	46.78116
60	0.37524	0.30487	0.68011	5.92961	4.08012	10.00973	5.92961	4.08012	10.00973	15.80218	13.38315	14.71781
Σ	–	–	–	33.37225	31.75775	65.13000	–	–	–	–	–	–

Table 5.2

The survivors, number of person-years, total after lifetime and life expectancies at the age x distributed by causes of death studied according to the abridged Hungarian life table by causes of death for females (1990)

Age (years) x	$l^{(F)}_{i,x}$	$l^{(F)}_{\neg i,x}$	$l^{(F)}_{x}$	$_nL^{(F)}_{i,x}$	$_nL^{(F)}_{\neg i,x}$	$_nL^{(F)}_{x}$	$T^{(F)}_{i,x}$	$T^{(F)}_{\neg i,x}$	$T^{(F)}_{x}$	$e^{0(F)}_{i,x}$	$e^{0(F)}_{\neg i,x}$	$e^{0(F)}_{x}$
(1)	(2)	(3)	(4)	(5)	(6)	(7)	(8)	(9)	(10)	(11)	(12)	(13)
0	0.58903	0.41097	1.00000	11.77467	7.90482	19.67949	46.08323	27.62687	73.71010	78.23579	67.22357	73.71010
20	0.58868	0.39199	0.98067	23.22074	14.59595	37.81669	34.30856	19.72205	54.03061	58.28049	50.31264	55.09561
60	0.55067	0.30179	0.85246	11.08782	5.12610	16.21392	11.08782	5.12610	16.21392	20.13514	16.98565	19.02015
Σ	–	–	–	46.08323	27.62687	73.71010	–	–	–	–	–	–

Table 6

Structure of survivors, i.e. of deceased at the age x and over by the groups of causes of death studied according to abridged Hungarian life tables by causes of death for males and females (1990)

Age (years) x	$l^{(M)}_{i,x}/l^{(M)}_x$	$l^{(M)}_{-i,x}/l^{(M)}_x$	$l^{(M)}_x/l^{(M)}_x$	$l^{(F)}_{i,x}/l^{(F)}_x$	$l^{(F)}_{-i,x}/l^{(F)}_x$	$l^{(F)}_x/l^{(F)}_x$
(1)	(2)	(3)	(4)	(5)	(6)	(7)
0	0.47224	0.52776	1.00000	0.58903	0.41097	1.00000
20	0.48466	0.51534	1.00000	0.60028	0.39972	1.00000
60	0.55173	0.44827	1.00000	0.64598	0.35402	1.00000

Table 7

The life expectancies at age x as weihted arithmetic means and the differences between them due to causes of death studied according to the abridged Hungarian male and female life tables by causes of death (1990)

Age (years) x	$e^{0(F)}_{i,x}(l^{(F)}_{i,x}/l^{(F)}_x)$	$e^{0(F)}_{-i,x}(l^{(F)}_{-i,x}/l^{(F)}_x)$	$e^{0(F)}_x$	$e^{0(M)}_{i,x}(l^{(M)}_{i,x}/l^{(M)}_x)$	$e^{0(M)}_{-i,x}(l^{(M)}_{-i,x}/l^{(M)}_x)$	$e^{0(M)}_x$	Differences due to		
							all causes of death	circulatory diseases	all other causes of death
(1)	(2)	(3)	(4)=(2)+(3)	(5)	(6)	(7)=(5)+(6)	(8)=(4)-(7)	(9)=(2)-(5)	(10)=(3)-(6)
0	46.08323	27.62687	73.71010	33.37225	31.75775	65.13000	8.58010	12.71098	-4.13088
20	34.98461	20.11097	55.09558	24.58781	22.19338	46.78119	8.31439	10.39680	-2.08241
60	13.00690	6.01326	19.02016	8.71854	5.99926	14.71780	4.30236	4.28836	0.01400

Table 8
Decomposition of differences between $_nd^{(F)}_x$ and $_nd^{(M)}_x$

$$_nd^{(F)}_x = {}_nd^{(M)}_x + \left[{}_nd^{(F)}_x - {}_nd^{(M)}_x \right] =$$

$$= {}_nd^{(M)}_x + \left[\left({}_nL^{(F)}_x - {}_nL^{(M)}_x \right) {}_nm^{(M)}_x + {}_nL^{(M)}_x \left({}_nm^{(F)}_x - {}_nm^{(M)}_x \right) + \left({}_nL^{(F)}_x - {}_nL^{(M)}_x \right) \left({}_nm^{(F)}_x - {}_nm^{(M)}_x \right) \right] =$$

$$= {}_nd^{(M)}_x + \left[\left({}_nL^{(F)}_x - {}_nL^{(M)}_x \right) {}_nm^{(M)}_x + {}_nL^{(M)}_x \left({}_nm^{(F)}_x - {}_nm^{(M)}_x \right) \right] + {}_nL^{(F)}_x \left({}_nm^{(F)}_x - {}_nm^{(M)}_x \right) \right]$$

Age groups (years) $x, x+n$	$_nd^{(M)}_x$	$\left({}_nL^{(F)}_x - {}_nL^{(M)}_x \right) {}_nm^{(M)}_x$	$_nL^{(M)}_x \left({}_nm^{(F)}_x - {}_nm^{(M)}_x \right)$	$\left({}_nL^{(F)}_x - {}_nL^{(M)}_x \right) \left({}_nm^{(F)}_x - {}_nm^{(M)}_x \right)$	$_nL^{(F)}_x \left({}_nm^{(F)}_x - {}_nm^{(M)}_x \right)$	$_nd^{(F)}_x - {}_nd^{(M)}_x$	$_nd^{(F)}_x$
(1)	(2)	(3)	(4)	(5)	(6)=(4)+(5)	(7)=(3)+(4)+ +(5)=(3)+(6)	(8)=(2)+(7)
0–19	0.02667	0.00014	-0.00745	-0.00003	-0.00748	-0.00734	0.01933
20–59	0.29322	0.01878	-0.17265	-0.01114	-0.18379	-0.16501	0.12821
60–	0.68011	0.42140	-0.15375	-0.09530	-0.24905	0.17235	0.85246
Σ	1.00000	0.44032	-0.33385	-0.10647	-0.44032	0.00000	1.00000

$$_nd_x^{(F)} = {}_nd_x^{(M)} + \left[{}_nd_x^{(F)} - {}_nd_x^{(M)}\right] =$$

$$= {}_nd_x^{(M)} + \left[(l_x^{(F)} - l_x^{(M)}) \cdot {}_nq_x^{(M)} + l_x^{(M)}({}_nq_x^{(F)} - {}_nq_x^{(M)}) + (l_x^{(F)} - l_x^{(M)})({}_nq_x^{(F)} - {}_nq_x^{(M)})\right] =$$

$$= {}_nd_x^{(M)} + \left[(l_x^{(F)} - l_x^{(M)}) \cdot {}_nq_x^{(M)} + l_x^{(F)}({}_nq_x^{(F)} - {}_nq_x^{(M)})\right]$$

Age groups (years) $x, x+n$	${}_nd_x^{(M)}$	$(l_x^{(F)}-l_x^{(M)}) \cdot {}_nq_x^{(M)}$	$l_x^{(M)}({}_nq_x^{(F)}-{}_nq_x^{(M)})$	$(l_x^{(F)}-l_x^{(M)})({}_nq_x^{(F)}-{}_nq_x^{(M)})$	$l_x^{(F)}({}_nq_x^{(F)}-{}_nq_x^{(M)})$	${}_nd_x^{(F)} - {}_nd_x^{(M)}$	${}_nd_x^{(F)}$
(1)	(2)	(3)	(4)	(5)	(6)=(4)+(5)	(7)=(3)+(4)+(5)=(3)+(6)	(8)=(2)+(7)
0–19	0.02667	0.00000	-0.00734	0.00000	-0.00734	-0.00734	0.01933
20–59	0.29322	0.00220	-0.16596	-0.00125	-0.16721	-0.16501	0.12821
60–	0.68011	0.17235	0.00000	0.00000	0.00000	0.17235	0.85246
Σ	1.00000	0.17455	-0.17330	-0.00125	-0.17455	0.00000	1.00000

Table 9
Decomposition of differences between $l^{(F)}_x$ and $l^{(M)}_x$

$$l^{(F)}_x = l^{(M)}_x + \left[l^{(F)}_x - l^{(M)}_x \right] =$$

$$= l^{(M)}_x + \left[\sum_\omega^x ({}_nL^{(F)}_x - {}_nL^{(M)}_x) \, {}_nm^{(M)}_x + \sum_\omega^x {}_nL^{(M)}_x ({}_nm^{(F)}_x - {}_nm^{(M)}_x) + \sum_\omega^x ({}_nL^{(F)}_x - {}_nL^{(M)}_x)({}_nm^{(F)}_x - {}_nm^{(M)}_x) \right] =$$

$$= l^{(M)}_x + \left[\sum_\omega^x ({}_nL^{(F)}_x - {}_nL^{(M)}_x) \, {}_nm^{(M)}_x + \sum_\omega^x {}_nL^{(F)}_x ({}_nm^{(F)}_x - {}_nm^{(M)}_x) \right].$$

Age groups (years) x	$l^{(M)}_x$	$\sum_\omega^x ({}_nL^{(F)}_x - {}_nL^{(M)}_x)\,{}_nm^{(M)}_x$	$\sum_\omega^x {}_nL^{(M)}_x ({}_nm^{(F)}_x - {}_nm^{(M)}_x)$	$\sum_\omega^x ({}_nL^{(F)}_x - {}_nL^{(M)}_x)({}_nm^{(F)}_x - {}_nm^{(M)}_x)$	$\sum_\omega^x {}_nL^{(F)}_x ({}_nm^{(F)}_x - {}_nm^{(M)}_x)$	$l^{(F)}_x - l^{(M)}_x$	$l^{(F)}_x$
(1)	(2)	(3)	(4)	(5)	(6)=(4)+(5)	(7)=(3)+(4)+(5)=(3)+(6)	(8)=(2)+(7)
0	1.00000	0.44032	−0.33385	−0.10647	−0.44032	0.00000	1.00000
20	0.97333	0.44018	−0.32640	−0.10644	−0.43284	0.00734	0.98067
60	0.68011	0.42140	−0.15375	−0.09530	−0.24905	0.17235	0.85246

Emil Valkovics

$$l_x^{(F)} = l_x^{(M)} + \left[l_x^{(F)} - l_x^{(M)}\right] =$$

$$= l_x^{(M)} + \left[\sum_\omega^x (l_x^{(F)} - l_x^{(M)}){}_nq_x^{(M)} + \sum_\omega^x l_x^{(M)}({}_nq_x^{(F)} - {}_nq_x^{(M)}) + \sum_\omega^x (l_x^{(F)} - l_x^{(M)})({}_nq_x^{(F)} - {}_nq_x^{(M)})\right] =$$

$$= l_x^{(M)} + \left[\sum_\omega^x (l_x^{(F)} - l_x^{(M)}){}_nq_x^{(M)} + \sum_\omega^x l_x^{(F)}({}_nq_x^{(F)} - {}_nq_x^{(M)})\right].$$

Age groups (years) x	$l_x^{(M)}$	$\sum_\omega^x (l_x^{(F)} - l_x^{(M)}){}_nq_x^{(M)}$	$\sum_\omega^x l_x^{(M)}({}_nq_x^{(F)} - {}_nq_x^{(M)})$	$\sum_\omega^x (l_x^{(F)} - l_x^{(M)})({}_nq_x^{(F)} - {}_nq_x^{(M)})$	$\sum_\omega^x l_x^{(F)}({}_nq_x^{(F)} - {}_nq_x^{(M)})$	$l_x^{(F)} - l_x^{(M)}$	$l_x^{(F)}$
(1)	(2)	(3)	(4)	(5)	(6)=(4)+(5)	(7)=(3)+(4)+ +(5)=(3)+(6)	(8)=(2)+(7)
0	1.00000	0.17455	−0.17330	−0.00125	−0.17455	0.00000	1.00000
20	0.97333	0.17455	−0.16596	−0.00125	−0.16721	0.00734	0.98067
60	0.68011	0.17235	0.00000	0.00000	0.00000	0.17235	0.85246

Table 10
Decomposition of differences between $T^{(F)}_x$ and $T^{(M)}_x$

$$T^{(F)}_x = T^{(M)}_x + \left[T^{(F)}_x - T^{(M)}_x\right] =$$

$$= T^{(M)}_x + \left[(l^{(F)}_x - l^{(M)}_x)e^{0(M)}_x + l^{(M)}_x(e^{0(F)}_x - e^{0(M)}_x) + (l^{(F)}_x - l^{(M)}_x)(e^{0(F)}_x - e^{0(M)}_x)\right] =$$

$$= T^{(M)}_x + \left[(l^{(F)}_x - l^{(M)}_x)e^{0(M)}_x + l^{(F)}_x(e^{0(F)}_x - e^{0(M)}_x)\right]$$

Age groups (years) x	$T^{(M)}_x$	$(l^{(F)}_x - l^{(M)}_x)e^{0(M)}_x$	$l^{(M)}_x(e^{0(F)}_x - e^{0(M)}_x)$	$(l^{(F)}_x - l^{(M)}_x)(e^{0(F)}_x - e^{0(M)}_x)$	$l^{(F)}_x(e^{0(F)}_x - e^{0(M)}_x)$	$T^{(F)}_x - T^{(M)}_x$	$T^{(F)}_x$
(1)	(2)	(3)	(4)	(5)	(6)=(4)+(5)	(7)=(3)+(4)+(5)=(3)+(6)	(8)=(2)+(7)
0	65.13000	0.00000	8.58010	0.00000	8.58010	8.58010	73.71010
20	45.53351	0.34337	8.09270	0.06103	8.15373	8.49710	54.03061
60	10.00973	2.53661	2.92606	0.74151	3.66757	6.20419	16.21392

Table 11
Decomposition of differences between $_nL^{(F)}_x$ and $_nL^{(M)}_x$

$$_nL^{(F)}_x = {}_nL^{(M)}_x + \left[{}_nL^{(F)}_x - {}_nL^{(M)}_x\right] =$$

$$= {}_nL^{(M)}_x + \left[\left(l^{(F)}_x - l^{(M)}_x\right)e^{0(M)}_x - \left(l^{(F)}_{x+n} - l^{(M)}_{x+n}\right)e^{0(M)}_{x+n}\right] + \left[l^{(M)}_x\left(e^{0(F)}_x - e^{0(M)}_x\right) - l^{(M)}_{x+n}\left(e^{0(F)}_{x+n} - e^{0(M)}_{x+n}\right)\right] +$$

$$+ \left[\left(l^{(F)}_x - l^{(M)}_x\right)\left(e^{0(F)}_x - e^{0(M)}_x\right) - \left(l^{(F)}_{x+n} - l^{(M)}_{x+n}\right)\left(e^{0(F)}_{x+n} - e^{0(M)}_{x+n}\right)\right] =$$

$$= {}_nL^{(M)}_x + \left[\left(l^{(F)}_x - l^{(M)}_x\right)e^{0(M)}_x - \left(l^{(F)}_{x+n} - l^{(M)}_{x+n}\right)e^{0(M)}_{x+n}\right] + \left[l^{(F)}_x\left(e^{0(F)}_x - e^{0(M)}_x\right) - l^{(F)}_{x+n}\left(e^{0(F)}_{x+n} - e^{0(M)}_{x+n}\right)\right]$$

Age groups (years) $x, x+n$	$_nL^{(M)}_x$	$(l^{(F)}_x - l^{(M)}_x)e^{0(M)}_x -$ $-(l^{(F)}_{x+n} - l^{(M)}_{x+n})e^{0(M)}_{x+n}$	$l^{(M)}_x(e^{0(F)}_x - e^{0(M)}_x) -$ $-l^{(M)}_{x+n}(e^{0(F)}_{x+n} - e^{0(M)}_{x+n})$	$(l^{(F)}_x - l^{(M)}_x)(e^{0(F)}_x - e^{0(M)}_x) -$ $-(l^{(F)}_{x+n} - l^{(M)}_{x+n})(e^{0(F)}_{x+n} - e^{0(M)}_{x+n})$	$l^{(F)}_x(e^{0(F)}_x - e^{0(M)}_x) -$ $-l^{(F)}_{x+n}(e^{0(F)}_{x+n} - e^{0(M)}_{x+n})$	$_nL^{(F)}_x - {}_nL^{(M)}_x$	$_nL^{(F)}_x$
(1)	(2)	(3)	(4)	(5)	(6)=(4)+(5)	(7)=(3)+(4)+ +(5)=(3)+(6)	(8)=(2)+(7)
0–19	19.59649	−0.34337	0.48740	−0.06103	0.42637	0.08300	19.67949
20–59	35.52378	−2.19324	5.16664	−0.68048	4.48616	2.29281	37.81669
60–	10.00973	2.53661	2.92606	0.74151	3.66757	6.20419	16.21392
Σ	65.13000	0.00000	8.58010	0.00000	8.58010	8.58010	73.71010

The column (6) contains the data we obtain when using the method of *Andreev*. Column (3) corrects the distribution presented by column (6) and column (5) corrects the distribution presented by column (3) and column (4).

Table 12

Contribution of mortality due to different groups of causes of death to the difference between the life expectancies at birth of Hungarian females and Hungarian males, 1990

Causes of death studied	ICD number 9th revision 1975	Life expectancies at birth (years)	Proportion of deceased	(3) × (4)	Life expectancies at birth (years)	Proportion of deceased	(6) × (7)	(8)−(5)
		in the case of males			in the case of females			
(1)	(2)	(3)	(4)	(5)	(6)	(7)	(8)	(9)
1. Infectious and parasitic diseases	001–139	61.476	0.00838	0.515	66.458	0.00448	0.298	−0.217
2. Neoplasms — all forms	140–239	65.960	0.22907	15.110	69.505	0.19164	13.320	−1.790
3. Diseases of the circulatory system	390–429, 440–459	68.692	0.26820	18.423	77.195	0.29453	22.736	4.313
4. Cerebrovascular diseases	430–438	73.266	0.20404	14.949	79.277	0.29450	23.347	8.398
5. Diseases of the respiratory system	460–519	69.398	0.05319	3.691	74.264	0.03777	2.805	−0.886
6. Diseases of the digestive system	520–579	58.952	0.07103	4.187	66.429	0.04900	3.255	−0.932
7. Congenital anomalies	740–759	8.918	0.00547	0.049	9.116	0.00448	0.041	−0.008
8. Certain conditions originating in the perinatal period	760–779	0.498	0.00948	0.005	0.499	0.00804	0.004	−0.001
9. Injury and poisoning	800–999	51.944	0.10954	5.690	68.753	0.06952	4.779	−0.911
10. All other causes of death	–	60.359	0.04160	2.511	67.872	0.04604	3.125	0.614
Total	–	–	1.00000	65.130	–	1.00000	73.710	8.580

List of Contributors

Andorka, Rudolf (1931–), sociologist, member of the Hungarian Academy of Sciences. From 1963 research fellow of the Demographic Research Institute, later of the Central Statistical Office. From 1984 professor of the University of Economics in Budapest. His main fields of research are social stratification and mobility, life styles, and social factors influencing the number of children in families.

Cseh-Szombathy, László (1925–), sociologist, member of the Hungarian Academy of Sciences. From 1973 university professor. Between 1982 and 1988 director of the Sociological Institute of the Academy. Since 1957 his main fields of reasearch have been suicide and the changing conditions of family life.

Csernák, Magdolna (1941–), Ph.D., lawyer and demographer. From 1970 research fellow of the Demographic Reseach Institute and from 1995 Director of the Institute. Secretary of the Demographic Committee of the Hungarian Academy of Sciences. Pursues research on the demographical aspects of family life, the foundation and dissolution of families, and the composition of families and households.

Hablicsek, László (1953–), mathematician and demographer. From 1978 research fellow of the Demographic Research Institute. His fields of research are population projection, demographic transitions, and aging of the population.

Józan, Péter (1935–), physician and demographer. Head of Department at the Central Statistical Office, member of the Demographic Committee of the Hungarian Academy of Sciences and the National Health Council. His fields of reseach are morbidity and mortality with special regard to chronic, non-infectious diseases and their consequences.

Kamarás, Ferenc (1943–), demographer. From 1971 research fellow of the Central Statistical Office working on various fields of population statistics. His main fields of reseach are fertility and family planning.

Miltényi, Károly (1930–), demographer and statistician. Between 1950 and 1990 he worked at the Central Statistical Office on social statistics and various demographical problems. Between 1990 and 1995 director of the Demographic Research Institute. Between 1965 and 1986 expert of the UNDP and the ILO in Ghana, Uganda, and Zambia. His fields of reseach are population policy and the economic and social aspects of demographical changes.

Pongrácz, Marietta (1944–), Ph.D., demographer. Research fellow, senior reseach fellow of the Demographic Research Institute since 1966. Her main fields of research are the demographic aspects of families, family policy, and fertility.

S. Molnár, Edit (1934–), Ph.D., sociologist. Scientific advisor of the Demographic Research Institute since 1973. Her main fields of research are founding of families, family policy, and public opinion research on demographic questions.

Tóth, Pál Péter (1942–), Ph.D., sociologist. Between 1975 and 1986 research fellow, associate professor at the Sociological Department of the Eötvös Lóránd University of Sciences. Later senior member of the Institute for Hungarian Studies, and from 1992 of the Demographic Research Institute. His main fields of research are international migration, and the sociological and demographic aspects of ethnic problems.

Valkovics, Emil (1930–), D.Sc., demographer. From 1958 research fellow of the Central Statistical Office, and from 1962 of the Demographic Research Institute. At present he is scientific adviser. His main fields of research are the principles and methods of demographic analysis, mathematical demography, multivariate statistical analysis, and the analysis of mortality by causes of death.

Volumes Published in
"Atlantic Studies on Society in Change"

A Series distributed by Columbia University Press

No. 1   *Tolerance and Movements of Religious Dissent in Eastern Europe.* Edited by Béla K. Király. 1977.

No. 2   *The Habsburg Empire in World War I.* Edited by R. A. Kann. 1978

No. 3   *The Mutual Effects of the Islamic and Judeo-Christian Worlds: The East European Pattern.* Edited by A. Ascher, T. Halasi-Kun, B. K. Király. 1979.

No. 4   *Before Watergate: Problems of Corruption in American Society.* Edited by A. S. Eisenstadt, A. Hoogenboom, H. L. Trefousse. 1979.

No. 5   *East Central European Perceptions of Early America.* Edited by B. K. Király and G. Bárány. 1977.

No. 6   *The Hungarian Revolution of 1956 in Retrospect.* Edited by B. K. Király and Paul Jónás. 1978.

No. 7   *Brooklyn U.S.A.: Fourth Largest City in America.* Edited by Rita S. Miller. 1979.

No. 8   *Prime Minister Gyula Andrássy's Influence on Habsburg Foreign Policy.* János Decsy. 1979.

No. 9   *The Great Impeacher: A Political Biography of James M. Ashley.* Robert F. Horowitz. 1979.

No. 10   *Special Topics and Generalizations on the Eighteenth and*
Vol. I*   *Nineteenth Century.* Edited by Béla K. Király and Gunther E. Rothenberg. 1979.

---

* Volumes Nos. I through XXXVI refer to the series *War and Society in East Central Europe.*

No. 11 Vol. II	*East Central European Society and War in the Pre-Revolutionary 18th-Century.* Edited by Gunther E. Rothenberg, Béla K. Király, and Peter F. Sugar. 1982.
No. 12 Vol. III	*From Hunyadi to Rákóczi: War and Society in Late Medieval and Early Modern Hungary.* Edited by János M. Bak and Béla K. Király. 1982.
No. 13 Vol. IV	*East Central European Society and War in the Era of Revolutions: 1775-1856.* Edited by B. K. Király. 1984.
No. 14 Vol. V	*Essays on World War I: Origins and Prisoners of War.* Edited by Samuel R. Williamson, Jr. and Peter Pastor. 1983.
No. 15 Vol. VI	*Essays on World War I: Total War and Peacemaking, A Case Study on Trianon.* Edited by B. K. Király, Peter Pastor, and Ivan Sanders. 1982.
No. 16 Vol. VII	*Army, Aristocracy, Monarchy: War, Society and Government in Austria, 1618-1780.* Edited by Thomas M. Barker. 1982.
No. 17 Vol. VIII	*The First Serbian Uprising 1804-1813.* Edited by Wayne S. Vucinich. 1982.
No. 18 Vol. IX	*Czechoslovak Policy and the Hungarian Minority 1945-1948.* Kálmán Janics. Edited by Stephen Borsody. 1982.
No. 19 Vol. X	*At the Brink of War and Peace: The Tito-Stalin Split in a Historic Perspective.* Edited by Wayne S. Vucinich. 1982.
No. 20	*Inflation Through the Ages: Economic, Social, Psychological and Historical Aspects.* Edited by Edward Marcus and Nathan Schmuckler. 1981.
No. 21	*Germany and America: Essays on Problems of International Relations and Immigration.* Edited by Hans L. Trefousse. 1980.
No. 22	*Brooklyn College: The First Half Century.* Murray M. Horowitz. 1981.
No. 23	*A New Deal for the World: Eleanor Roosevelt and American Foreign Policy.* Jason Berger. 1981.

No. 24	*The Legacy of Jewish Migration: 1881 and Its Impact.* Edited by David Berger. 1982.
No. 25	*The Road to Bellapais: Cypriot Exodus to Northern Cyprus.* Pierre Oberling. 1982.
No. 26	*New Hungarian Peasants: An East Central European Experience with Collectivization.* Edited by Marida Hollos and Béla C. Maday. 1983.
No. 27	*Germans in America: Aspects of German-American Relations in the Nineteenth Century.* Edited by Allen McCormick. 1983.
No. 28	*A Question of Empire: Leopold I and the War of Spanish Succession, 1701-1705.* Linda and Marsha Frey. 1983.
No. 29	*The Beginning of Cyrillic Printing — Cracow, 1491. From the Orthodox Past in Poland.* Szczepan K. Zimmer. Edited by Ludwik Krzyzanowski and Irene Nagurski. 1983.
No. 29a	*A Grand Ecole for the Grand Corps: The Recruitment and Training of the French Administration.* Thomas R. Osborne. 1983.
No. 30 Vol. XI	*The First War between Socialist States: The Hungarian Revolution of 1956 and Its Impact.* Edited by Béla K. Király, Barbara Lotze, Nandor Dreisziger. 1984.
No. 31 Vol. XII	*The Effects of World War I, The Uprooted: Hungarian Refugees and Their Impact on Hungary's Domestic Politics.* István Mócsy. 1983.
No. 32 Vol. XIII	*The Effects of World War I: The Class War after the Great War: The Rise Of Communist Parties in East Central Europe, 1918-1921.* Edited by Ivo Banac. 1983.
No. 33 Vol. XIV	*The Crucial Decade: East Central European Society and National Defense, 1859-1870.* Edited by Béla K. Király. 1984.
No. 35 Vol. XVI	*Effects of World War I: War Communism in Hungary, 1919.* György Péteri. 1984.
No. 36 Vol. XVII	*Insurrections, Wars, and the Eastern Crisis in the 1870s.* Edited by B. K. Király and Gale Stokes. 1985.

No. 37 Vol. XVIII	*East Central European Society and the Balkan Wars, 1912-1913.* Edited by B. K. Király and Dimitrije Djordjevic. 1986.
No. 38 Vol. XIX	*East Central European Society in World War I.* Edited by B. K. Király and N. F. Dreisziger, Assistant Editor Albert A. Nofi. 1985.
No. 39 Vol. XX	*Revolutions and Interventions in Hungary and Its Neighbor States, 1918-1919.* Edited by Peter Pastor. 1988.
No. 40 Vol. XXI	*East Central European Society and War, 1750-1920. Bibliography and Historiography.* Complied and edited by László Alföldi. Pending.
No. 41 Vol. XXII	*Essays on East Central European Society and War, 1740-1920.* Edited by Stephen Fischer-Galati and Béla K. Király. 1988.
No. 42 Vol. XXIII	*East Central European Maritime Commerce and Naval Policies, 1789-1913.* Edited by Apostolos E. Vacalopoulos, Constantinos D. Svolopoulos, and Béla K. Király. 1988.
No. 43 Vol. XXIV	*Selections, Social Origins, Education and Training of East Central European Officers Corps.* Edited by Béla K. Király and Walter Scott Dillard. 1988.
No. 44 Vol. XXV	*East Central European War Leaders: Civilian and Military.* Edited by Béla K. Király and Albert Nofi. 1988.
No. 46	*Germany's International Monetary Policy and the European Monetary System.* Hugo Kaufmann. 1985.
No. 47	*Iran Since the Revolution — Internal Dynamics, Regional Conflicts and the Superpowers.* Edited by Barry M. Rosen. 1985.
No. 48 Vol. XXVII	*The Press During the Hungarian Revolution of 1848-1849.* Domokos Kosáry. 1986.
No. 49	*The Spanish Inquisition and the Inquisitional Mind.* Edited by Angel Alcala. 1987.
No. 50	*Catholics, the State and the European Radical Right, 1919-1945.* Edited by Richard Wolff and Jorg K. Hoensch. 1987.

No. 51 Vol. XXVIII	*The Boer War and Military Reforms.* Jay Stone and Erwin A. Schmidl. 1987.
No. 52	*Baron Joseph Eötvös, A Literary Biography.* Steven B. Várdy. 1987.
No. 53	*Towards the Renaissance of Puerto Rican Studies: Ethnic and Area Studies in University Education.* Maria Sanchez and Antonio M. Stevens. 1987.
No. 54	*The Brazilian Diamonds in Contracts, Contraband and Capital.* Harry Bernstein. 1987.
No. 55	*Christians, Jews and Other Worlds: Patterns of Conflict and Accommodation.* Edited by Phillip F. Gallagher. 1988.
No. 56 Vol. XXVI	*The Fall of the Medieval Kingdom of Hungary: Mohács, 1526, Buda, 1541.* Géza Perjés. 1989.
No. 57	*The Lord Mayor of Lisbon: The Portuguese Tribune of the People and His 24 Guilds.* Harry Bernstein. 1989.
No. 58	*Hungarian Statesmen of Destiny: 1860-1960.* Edited by Paul Böbdy. 1989.
No. 59	*For China: The Memoirs of T. G. Li, former Major General in the Chinese Nationist Army.* T. G. Li. Written in collaboration with Roman Rome. 1989.
No. 60	*Politics in Hungary: For A Democratic Alternative.* János Kis, with an Introduction by Timothy Garton Ash. 1989.
No. 61	*Hungarian Worker's Councils in 1956.* Edited by Bill Lomax. 1990.
No. 62	*Essays on the Structure and Reform of Centrally Planned Economic Systems.* Paul Jonas. A joint publication with Corvina Kiadó, Budapest. 1990.
No. 63	*Kossuth as a Journalist in England.* Éva H. Haraszti. A joint publication with Akadémiai Kiadó, Budapest. 1990.
No. 64	*From Padua to the Trianon, 1918-1920.* Mária Ormos. A joint publication with Akadémiai Kiadó, Budapest. 1990.
No. 65	*Towns in Medieval Hungary.* Edited by László Gerevich. A joint publication with Akadémiai Kiadó, Budapest. 1990.

No. 66   *The Nationalities Problem in Transylvania, 1867-1940.* Sándor Bíró. 1992.

No. 67   *Hungarian Exiles and the Romanian National Movement, 1849-1867.* Béla Borsi-Kálmán. 1991.

No. 68   *The Hungarian Minority's Situation in Ceausescu's Romania.* Edited by Rudolf Joó and Andrew Ludanyi. 1994.

No. 69   *Democracy, Revolution, Self-Determination. Selected Writings.* István Bibó. Edited by Károly Nagy. 1991.

No. 70   *Trianon and the Protection of Minorities.* József Galántai. A joint publication with Corvina Kiadó, Budapest. 1991.

No. 71   *King Saint Stephen of Hungary.* György Györffy. 1994.

No. 72   *Dynasty, Politics and Culture. Selected Essays.* Robert A. Kann. Edited by Stanley B. Winters. 1991.

No. 73   *Jadwiga of Anjou and the Rise of East Central Europe.* Oscar Halecki. Edited by Thaddeus V. Gromada. A joint publication with the Polish Institute of Arts and Sciences of America, New York. 1991.

No. 74
Vol. XXIX   *Hungarian Economy and Society During World War Two.* Edited by György Lengyel. 1993.

No. 75   *The Life of a Communist Revolutionary, Béla Kun.* György Borsányi. 1993.

No. 76   *Yugoslavia: The Process of Disintegration.* Laslo Sekelj. 1993.

No. 77
Vol. XXX   *Wartime American Plans for a New Hungary. Documents from the U.S. Department of State, 1942-1944.* Edited by Ignác Romsics. 1992.

No.78
Vol. XXXI   *Planning for War against Russia and Serbia. Austro-Hungarian and German Military Strategies, 1871-1914.* Graydon A. Tunstall, Jr. 1993.

No. 79   *American Effects on Hungarian Imagination and Political Thought, 1559-1848.* Géza Závodszky. 1995.

No. 80
Vol. XXXII   *Trianon and East Central Europe: Antecedents and Repercussions.* Edited by Béla K. Király and László Veszprémy. 1995.

No. 81	*Hungarians and Their Neighbors in Modern Times, 1867-1950.* Edited by Ferenc Glatz. 1995.
No. 82	*István Bethlen: A Great Conservative Statesman of Hungary, 1874-1946.* Ignác Romsics. 1995.
No. 83 Vol. XXXIII	*20th Century Hungary and the Great Powers.* Edited by Ignác Romsics. 1995.
No. 84	*Lawful Revolution in Hungary, 1989-1994.* Edited by Béla K. Király and András Bozóki. 1995.
No. 85	*The Demography of Contemporary Hungarian Society.* Edited by Pál Péter Tóth and Emil Valkovich. 1996.
No. 86	*Budapest, A History from its Beginnings to 1996.* Edited by András Gerõ and János Poór. 1996.
No. 87	*József Eötvös's "The Dominant Ideas of the Nineteenth Century and their Impact on the State."* Volume 1. Diagnosis. Translated, edited, anootated and indexed with an introductory essay by D. Mervyn Jones. 1996.
No. 88	*József Eötvös's "The Dominant Ideas of the Nineteenth Century and their Impact on the State."* Volume 2. Remedy. Translated, edited, anootated and indexed with an introductory essay by D. Mervyn Jones. 1997.
No. 89	*The Social History of the Hungarian Intelligentsia's in the "Long Nineteenth Century,"* 1825–1914. János Mazsu. 1996.
No. 90 Vol. XXXIV	*Pax Britannica: Wartime Foreign Office Documents Regarding Plans for a Post Bellum East Central Europe.* Edited by András D. Bán. 1997.
No. 91	*National Identity in Contemporary Hungary.* György Csepeli. 1996.
No. 92	*The Hungarian Parliament, 1867–1918: A Mirage of Power.* András Gerõ. 1997.
No. 93 Vol. XXXV	*The Hungarian Revolution and War for Independence, 1848–1849.* A Military History. Edited by Gábor Bona. 1997.
No. 94	*The End of Assimilation: "The Jewish Question" in Hungary.* Tamás Ungvári. 1997.

No. 95	*Academia and State Socialism: Essays on the Political History of Academic Life in Post-1945 Hungary and East Central Europe.* György Péteri. 1997.
No. 96 Vol. XXXVI	*Through the Prism of the Habsburg Monarchy: Hungary in American Diplomacy and Public Opinion during World War I.* Tibor Glant. 1997.